Weights and Measures Program Requirements

A Handbook for the Weights and Measures Administrator

Authors:

Carol Hockert, Chief
Henry V. Oppermann

National Institute of Standards and Technology
Weights and Measures Division
Gaithersburg, MD 20899-2600

U. S. Department of Commerce
Gary Locke, Secretary

**National Institute of
Standards and Technology**
Patrick D. Gallagher, Director

NIST
Handbook **155**

2011

Certain commercial entities, equipment, or materials may be identified in this document in order to describe an experimental procedure or concept adequately. Such identification is not intended to imply recommendation or endorsement by the National Institute of Standards and Technology, nor is it intended to imply that the entities, materials, or equipment are necessarily the best available for the purpose.

National Institute of Standards and Technology Handbook 155, 2011 Edition
Natl. Inst. Stand. Technol. Handb. 155, 2011 Ed., 91 pages (March 2011)

Washington: 2011

Preface

When a weights and measures administrator makes decisions within a specific jurisdiction, it is beneficial to understand the scope of the entire system and to reflect upon methods and practices that have been tested over the years. This handbook was developed for the weights and measures administrator to be used as a reference tool. It is designed to read in sections as needed for a specific situation or to learn piece by piece about the weights and measures system as a whole. The reader is invited to submit comments on any section within this document and to provide updates and corrections as they are identified.

Table of Contents

List of Tables and Figures

Weights and Measures Program Requirements and Assessment

1.0   The Commercial Measurement System

Many commercial transactions are based on weight, volume, length, or count of products bought and sold. Packaged goods are purchased at the supermarket, people buy delicatessen items over price computing scales, gasoline and diesel fuel are purchased through pumps (retail motor fuel dispensers), gasoline and diesel fuel must meet prescribed quality or octane standards, scanners are used at checkout stands in retail stores to look up prices of products identified by bar codes, farmers sell grain, produce, and livestock over scales, grain prices are adjusted up or down based upon quality measurements, and landfills charge fees based upon the weight of the trash delivered. The structure within which transactions among businesses and with the general public are conducted is called the commercial measurement system.

Weights and measures activities are pervasive within the United States. It is estimated that U.S. weights and measures regulations impact roughly half of the U.S. gross domestic product. The success of the commercial measurement system can be judged by the ease with which transactions are executed, the level of confidence that buyers and sellers have, and the accuracy with which these transactions are performed.

In a well-functioning commercial measurement system, effective laws and regulations are in place to ensure an orderly marketplace. The laws and regulations should provide consumer protection by preventing deceptive and misleading practices, but should not be overly burdensome to businesses. They should also foster fair competition among companies in the many different facets of the commercial measurement system. Finally, the laws, regulations, and technical standards must be sufficiently flexible to adjust to new technology and marketing practices. Determining the correct balance of these many factors is a major and ongoing challenge to the weights and measures community.

The weights and measures community includes a wide range of organizations and functions:

- Businesses that sell to the public, manufacturers of the measuring instruments (scales and metering devices) used in direct sale and in the packaging of goods;

- Manufacturers that package the wide variety of goods available to consumers, producers of the raw materials and food products that go into the consumer goods;

- Raw materials used by companies in the chain of manufacturing and production that lead to the final consumer products; and

- The weights and measures programs that regulate the marketplace so equity exists in all transactions.

- Federal agencies that have regulatory jurisdiction over most products.

1

The activities of weights and measures regulatory programs are often invisible to the public. Generally, the public sees only the transactions that occur in the marketplace, such as the weighing of meat in the delicatessen, the weighing of produce at the checkout counter and the dispensing of fuel into their automobiles in the service stations. The complex infrastructure and the many activities involved in implementing the programs are not apparent to consumers, and a well functioning marketplace is often taken for granted. This reality makes it important to make ongoing efforts to educate the public and legislators, to help ensure support for the programs.

The delivery of full weight and measure and the elimination of fraud and misrepresentation have been objectives in commercial transactions from the time of the inception of quantity determination of merchandise down to the present day. It has been demonstrated that there are always some who will avail themselves of an opportunity for an unfair or dishonest advantage, and that, even though this number be relatively small, the results of their fraudulent practices constitute a serious problem in their community. Again, it has been shown that another group, larger than the one just mentioned but still constituting only a small percentage of those engaged in business, are careless in the conduct of their affairs to such a degree that the community suffers almost as much from their unintentional errors as from the intentional inaccuracies of the fraudulently minded. Still a third group adds its share to the total of inequities attendant upon commercial quantity determination, and this is made up of those whose errors result from ignorance rather than from carelessness or intent to defraud. Of these three groups, one can be more sympathetic toward the last, the ones who know no better, than toward the other two. But it must not be overlooked that short weight or measure is equally damaging to the injured party whatever its underlying cause.

For the most part, transactions are accurate. Most merchants operate in good faith and are honest. However, oversight of the commercial measurement system is essential. Responsibility for oversight is shared among the federal, state, and local governments. The bulk of the weights and measures enforcement responsibilities reside in the state and local jurisdictions, although some federal agencies have been given specific weights and measures authority in some areas.

In the United States, "weights and measures" commonly refers to the infrastructure that supports the "commercial measurement system," and is frequently interchanged with "legal metrology". Internationally, the term "legal metrology" also includes the measurements for the medical fields, monitoring environmental pollution, acoustics, ionizing radiation, blood alcohol measurements, and other areas.

The weights and measures infrastructure includes the following:

- Standards and Units. The internationally defined units of measurement and the intrinsic and physical standards used as the basis for measurement in the any measurement system.

- Laws, Regulations, and Practices. The development of the standards, laws, and regulations that define the parameters for the quantity and quality of products sold among businesses and to the public, the methods of sale and information disclosure in transactions that permit consumers to make value comparisons and the practices for ensuring accurate weighing and measuring operations.

2

- Metrology Laboratories. The metrology laboratories transfer values and uncertainties from laboratory standards to lower level standards to establish and maintain a chain of traceability of physical standards to the international standards and the International System of Units[1] (SI).

- Weighing and Measuring Devices. The manufacture and approval of weighing and measuring devices used to determine the quantity of products, the cost of services based on quantities, or the quality of products (such as gasoline octane).

- Packaged Products. The processing and packaging of standard and random weight, volume, count, and length products for sale in stores.

- Service Industries. The service industries that maintain the compliance of weighing and measuring devices used in commercial transactions, applying accuracy and specification requirements.

- Weights and Measures Regulation. The regulatory oversight of the quantities, qualities, and cost of services based on measurements that form the basis of sale of products and services. This oversight includes inspection procedures for ensuring device accuracy and use, and for verifying the net contents of packaged goods.

- Commercial Transactions. The transactions that are the basis of the transfer of goods from one party to another or the charging for services based on measurements.

The components of the weights and measures infrastructure help to ensure the accuracy and validity of commercial transactions based upon weight, measure, or count and to ensure that the quality of products meets required quality standards. Another purpose of these components is to ensure consumers are informed so that they can make value comparisons. A robust infrastructure ensures equity in the marketplace, meaning that consumers receive the correct quantity and quality of products and services for which they pay and businesses receive fair payment for the products and services that they deliver. By ensuring that they operate according to a consistent set of weights and measures standards and practices, businesses are also protected from unfair competition.

2.0   Weights and Measures Laws and Regulations

The power, authority and responsibilities of the weights and measures regulatory program must be clearly defined in the weights and measures law and is a critical basis for regulatory actions. The weights and measures statute is the foundation upon which the structure of weights and measures supervision is based. Without an adequate foundation no enduring building can be erected; likewise, without an adequate law for a basis it is impossible to erect a comprehensive system of weights and measures supervision that will successfully resist the unexpected changes of today's complicated and strenuous commercial life, or to realize the full measure of protection

---

[1] The International System of Units (SI), commonly known as the metric system, is the international standard for measurement. SI is made up of 7 base units that define the 22 derived units with special names and symbols.

that such a system should afford. It is of primary importance, therefore, that this basic law be carefully planned along broad lines to meet the urgent needs of business, that it be executed with precision and with an attention to detail that will insure a cohesive and substantial whole, and that it be reinforced by such provisions for administrative authority and penalties for violations of its provisions as will make possible effective enforcement.

The weights and measures law should not be so specific such that it restricts innovation of technology or marketing practices. Consequently, while the implementation of regulations is needed to provide structure for the marketplace, the revision and update of the regulations are also essential to provide the flexibility to respond to changes in technology and evolution of the marketplace. For example, e-business is now a major activity, since the use the Internet has become commonplace.

Important references for the development or updating of weights and measures laws are the National Institute of Standards and Technology (NIST) Handbook 130, *Uniform Laws and Regulations in the Areas of Legal Metrology and Engine Fuel Quality* and International Organization of Legal Metrology[2] (OIML) Document 1 "Elements for a Law on Legal Metrology." The weights and measures law should provide the authority for the weights and measures director to issue regulations to supplement the weights and measures laws. Below are some major components of a weights and measures law that should be included and several important regulations that should be developed to supplement the law. Table 1 shows components of laws and a list of supplementary laws and regulations.

---

[2] The OIML is an intergovernmental treaty organization whose membership includes Member States, countries which participate actively in technical activities, and Corresponding Members, courtiers which join the OIML as observers. It was established in 1955 in order to promote the global harmonization of legal metrology procedures. See http://www.oiml/org/about/presentation.html.

Table 1. Components of Weights and Measurements Laws

Components of a Weights and Measures Law	Supplementary Laws and Regulations
• Definitions • Legal Units of Measurement • Physical Standards • Technical Requirements for Measuring Instruments • Requirements for Type Evaluation • Responsibilities of the Weights and Measures Agency • Authority to Issue Regulations • Powers and Duties of the Director • Enforcement Authority • Misrepresentation of Quantity • Method of Sale • Sale from Bulk • Information Required on Packages • Declarations of Unit Price on Random Weight Packages • Advertising Packages for Sale • Civil Penalties • Criminal Penalties • Financial Provisions • Restraining Order and Injunction • Presumptive Evidence • Separability Provision • Repeal of Conflicting Laws • Regulations to be Unaffected by Repeal of Prior Enabling Statute • Effective Date	• Weighmaster law • Engine fuels, petroleum products, and automotive lubricants inspection law • Technical regulations for measuring instruments • Type evaluation for measuring instruments • Net contents of packaged goods • Method of sale regulation • Packaging and labeling regulation • Unit pricing regulation • Open dating regulation • Voluntary or mandatory registration of service agencies regulation

Weights and measures laws, regulations, test procedures and interpretations of weights and measures requirements must be consistent with national standards and recommendations. States should strive to follow the uniform laws and regulations contained in NIST Handbook 130, which contains the recommendations adopted by the National Conference on Weights and Measures (NCWM). Following the national standards is not necessarily a simple matter because the priorities of weights and measures programs must reflect the economic interests of the individual states.

Funding for the weights and measures regulatory programs is provided by the individual states based upon the priorities of the governors, the state legislatures, and the agency administrators. Sometimes the political objectives of elected or appointed officials are not consistent with the regulatory goals and responsibilities of the weights and measures program. These discrepancies in objectives may lead to differences among weights and measures programs. The need for international consistency in weights and measures requirements as a result of global manufacturing and marketing further complicates the picture. Nevertheless, the commitment to national and international uniformity must be pursued.

The application and implementation of the requirements must also be consistent across the states. Inconsistencies in weights and measures requirements can cause problems and frustrations for retailers, packagers, and instrument manufacturers who market nationally and must attempt to comply with inconsistent requirements. These companies market on a regional or national basis. Deviations disrupt their operations, complicate their efforts to comply with varying requirements, and increase the cost of products.

Variations from the uniform laws and regulations have the potential to disrupt interstate commerce. Consequently, it is recommended that each state document any variations that they have and make them available to companies that do business within their state. While this documentation is not required, it is a way to inform businesses of any unique requirements that the state may have.

Keeping businesses informed of these variations will increase awareness of any unique requirements, and they will therefore be in a better position to comply. If the state weights and measures program documents these variations, then each individual business does not have to do its own research to identify the unique requirements and possibly overlook some important requirement. Weights and measures programs benefit when businesses act on their own initiative (voluntary compliance) to comply with weights and measures requirements rather than relying on weights and measures officials to educate each business on the unique requirements as an ongoing part of inspections.

The inspection procedures for price verification, the interpretation of method of sale requirements, and test procedures for engine and heating fuels should be consistent with national recommendations and interpretations. The retailers, product manufacturers (packagers), and manufacturers of measuring instruments are often the best sources of information as to which states and for which requirements those states vary from the national standards and recommendations, since they are the ones that have the problem of complying with different or conflicting requirements. Weights and measures directors should consult with these businesses to learn if their state has unique practices or policies that differ from national recommendations or interpretations and which create a burden on business.

The test procedures and interpretation of the technical requirements for the inspection and test of measuring instruments must be consistent among the weights and measures programs. If instrument manufacturers must design scales and metering devices to meet different requirements, costs increase for both hardware and software development. State programs that deviate from the national standards should consider the value of any local differences in weights and measures requirements. Differences among state weights and measures requirements are obvious and serious obstacles to uniformity in weights and measures.

The procedures in NIST Handbook 133, *Checking the Net Content of Packaged Goods*, should be used by all weights and measures officials for checking the net content of packages. Additionally, weights and measures officials should pool their inspection results for products and companies to obtain a better assessment of a packager's packaging process regarding net content. The process for developing and updating weights and measures laws and technical regulations should be sufficiently responsive to keep up with changing technology and marketing practices to the extent that the new technologies have comparable accuracy as existing technologies and new marketing practices are not deceptive or confusing to consumers. Consumer preferences change over time and the marketplace changes in response to consumer priorities and niche markets. Weights and measures requirements should not be impediments to the development of new measurement technologies or new marketing practices that do not deceive the consumer.

Participation in the regional weights and measures associations and the National Conference on Weights and Measures are indications that each weights and measures program is acting to keep current with changes in the marketplace, updating technical requirements in response to changing technologies and markets and making an effort to keep current and educated on new issues and problems.

To progress toward uniformity, weights and measures officials must be knowledgeable on new technology and emerging issues, coordinate with their counterparts in other states and implement the decisions of the NCWM. The commercial marketplace, measurement instruments, and weights and measures enforcement have become too complicated for states to operate in isolation. A unified, nationally coordinated effort is needed to make weights and measures enforcement more efficient and effective.

The globalization of manufacturing, marketing and distribution has increased competition among companies around the world. In order to remain competitive and to grow, companies have had to merge and expand to cut costs and take advantage of global resources to become more efficient. Some companies that were able to thrive serving the U.S. market have found that they must expand internationally to grow. Consequently, many more companies market internationally and now support the use of international legal metrology standards for all countries.

The globalization of the marketplace is driving weights and measures requirements toward the international standards. The common mantra among many industry sectors is one standard, one test and accepted throughout the world. Therefore, many companies encourage the adoption of the OIML Recommendations and they want one type evaluation of measuring instruments to be accepted throughout the world. The United States and the NCWM are being pressured to more closely join the international effort to have compatible technical regulations in legal metrology. Significant steps have been made in the United States toward harmonization, including the new scale code established in the 1980s and the Mutual Recognition Agreement (MRA) with Canada. For many years, the NCWM committees have considered OIML recommendations as the basis for U.S. requirements. International standards have also been harmonized with those of the United States, indicating the reciprocal nature of this process. The NCWM is an active participating member of the OIML Mutual Acceptance Arrangement (MAA) for the type evaluation of certain measuring instruments.

3.0   The Regulatory Function of Weights and Measures

The primary function of the weights and measures official is to see to it that equity prevails in all commercial transactions involving determinations of quantity. As a regulatory official, the weights and measures inspector is an independent and objective third party to see that the interests of both the buyer and the seller are safeguarded. The delivery of full weight and measure and the elimination of fraud and misrepresentation have been objectives in commercial transactions from the time of the inception of quantity determination of merchandise down to the present day.

The objectives of legal metrology oversight include the following:

- Ensure the accuracy of commercial transactions;

- Ensure that commercial weighing and measuring devices comply with legal metrology requirements;

- Provide consumer protection;

- Ensure fair competition among businesses;

- Facilitate value comparisons by consumers; and

- Facilitate commerce and international trade.

Regulatory control must be exercised in an efficient and effective manner. The marketplace changes continually, so weights and measures approaches to regulatory oversight must also change. For example, in the early 1900s, most retail sales were from bulk in the form of direct sales to consumers. Now, a large percentage of transactions are for goods contained in packaged form.

In the early 1900s, weights and measures inspections also focused on the testing of the measuring instruments, because such a large percentage of commerce occurred over the measuring instruments. In some segments of the market, this remains the case. For example, sales of gasoline and diesel fuel for cars and trucks are direct sales. The gasoline and diesel fuel are measured as the fuel goes into the vehicle tank. The consumer has no way to verify the accuracy of the transaction and must rely on the accuracy of the fuel dispenser. For this reason, weights and measures officials regularly inspect and test fuel dispensers.

Similarly, farmers sell grain and produce over vehicle scales, and in most cases, measurement on the vehicle scale is the only time when the product is weighed. It is usually time consuming and impractical for farmers (and the weights and measures official) to get a second weight on trucks to verify the accuracy of the transactions. Consequently, most weights and measures programs put considerable resources into the testing of vehicle scales.

In direct sales, it is critical that the scale or meter be within legal tolerances in order to conduct an accurate transaction. The tolerances for measuring instruments are established such that the "Tolerances values are so fixed that the permissible errors are sufficiently small that there is no serious injury to either the buyer or the seller of commodities, yet not so small as to make manufacturing or maintenance costs of equipment disproportionately high."[3]

An accurate measuring instrument by itself does not ensure an accurate transaction. If the tare weight is incorrect, then the net weight will be inaccurate even if the scale is accurate. For example, in many states, businesses are allowed to use stored tare weights for trucks that are delivering sand, gravel and even grain. These tare weights vary because of such factors as the amount of fuel in the fuel tanks, the number of people in the cab of the truck, debris buildup, or replaced tires and equipment. Many times the stored tare weights are used for extended periods of time without updating the tare weights. Wind, rain, ice and snow may also affect the result of the weighing process and, therefore, affect the accuracy of the transaction. These errors in the tare weights may be many times larger than the tolerance that is permitted on the vehicle scale on which the trucks are weighed.

Weights and measures officials also must be alert for fraudulent activities. The weighing or measuring process must be performed properly in addition to having an accurate device to obtain an accurate transaction. Consequently, weights and measures officials must do much more than simply check the accuracy of measuring instruments. Unscrupulous business people may use a device fraudulently in an effort to cheat consumers. Weights and measures officials have also uncovered the unauthorized and fraudulent modification of the manufacturer software programmed in gasoline dispensers.

Undercover investigations of possible fraudulent practices and consumer complaints are resource-intensive activities. Therefore, weights and measures program managers must balance the allocation of resources to different activities to best maintain equity in the marketplace. No weights and measures program is funded at the level where the program can conduct regular oversight of all marketplace transactions. Therefore, most administrators focus on areas where their efforts have the greatest impact.

Many of the specifications contained in NIST Handbook 44, *Specifications, Tolerances, and Other Technical Requirements for Weighing and Measuring Devices,* are written to reduce the potential for fraudulent manipulation and use of the measuring instruments. In a direct sale situation, Handbook 44 requires that transactions take place in a manner that the consumer can observe the weighing or measuring operation and that specific information be provided to the consumer so that the consumer can verify the critical values of the transaction.

In the case of packaged goods, the consumer usually does not have the equipment or knowledge to verify the accuracy of the stated net contents. Furthermore, packaged goods are sold on the

---

[3] NIST Handbook 44, 2011 Edition, *Specifications, Tolerances, and Other Technical Requirements for Weighing and Measuring Devices,* Appendix A – Fundamental Considerations, page A-4, Section 2.2. Theory of Tolerances.

basis of average net content, with maximum allowable variation[4] (MAV) specified for individual packages. Consideration must also be given to moisture loss during good storage and distribution of packages. Consequently, the consumer must rely on the packager and the oversight of the weights and measures regulatory official to ensure that packaged goods meet the average and maximum allowable variation requirements.

Many state and local jurisdictions expanded their package inspection programs in the 1960s and 1970s to reflect the shift from the direct sale of bulk items to prepackaged consumer goods. This time period had a significant focus on consumer protection. The inspection of prepackaged goods requires specialized test equipment and the test methods can be time consuming. In addition, some products lose weight due to moisture loss, which must be recognized by officials. This may require additional inspections of some products in order to ascertain whether the packages are underweight due to reasonable moisture loss or poor package filling practices.

Despite the complicated nature of net content inspections, inspection activities in these areas should be conducted. Efforts must be made to ensure that valid sampling and inspection procedures are used and results shared with other weights and measures programs to ensure that effective regulatory oversight is exercised over all commercial transactions.

4.0   The Complexity of Weights and Measures Regulation

The legal metrology system is complex due to its range of transactions, products, measurements, and devices. Weights and measures officials have the responsibility and authority to enforce the accuracy of transactions among businesses as well as sales to and purchases from consumers. The marketplace is continually changing and is truly global. For many U.S. companies, a major growth opportunity is exporting, which, in turn, supports the U.S. economy. The international market is causing a convergence in weights and measures requirements, so U.S. weights and measures officials must be involved in the development of international standards and support international activities if they wish to influence the international standards and infrastructure. The range of international legal metrology topics is much broader than the scope of legal metrology issues addressed by weights and measures officials in the United States.

Measurements of quantity and quality are the foundation for efficient manufacturing and accurate transactions among businesses and to consumers. Competition forces companies to control variables affecting the quantity and quality of the products produced and to reduce waste in the manufacturing process. Efficient manufacturing processes incorporate accurate weighing and measuring devices into production and distribution processes to ensure effective control of the production variables. Over the years, the responsibilities of weights and measures officials have expanded to include the verification of quality statements and measurements in areas such as retail motor fuels, grain moisture, and protein measurements.

The bulk of the weights and measures regulatory authority is the responsibility of the states and most weights and measures laws and regulations are adopted at the state and local level. In a few

---

[4] Maximum Allowable Variation (MAV) is a deficiency in weight, measure, or count of an individual package beyond which the deficiency is considered to be an "unreasonable error." The number of packages with deficiencies that are greater than the MAV is controlled by the sampling procedure.

cases, Congress has given regulatory authority to Federal agencies. The U.S. Department of Agriculture (USDA) has regulatory authority regarding meat and poultry products, and grain transactions for export. The Food and Drug Administration (FDA) controls the labeling of many foods and pharmaceutical products. The Federal Trade Commission (FTC) also regulates labeling and advertising. NIST is not a regulatory agency. Congress has charged NIST with the responsibility to define the units of weights and measures and to work with the states to secure uniformity in weights and measures requirements and procedures to facilitate trade, both nationally and internationally.

International (and regional) trade agreements, such as the World Trade Organization Agreement on Technical Barriers to Trade, obligate the signatory countries to eliminate technical barriers to trade. The globalization of the marketplace has driven multinational companies to support the use of international standards as the basis for trade and regulations. The United States became a member of OIML in 1972. Consequently, the United States has a moral obligation to adopt OIML standards to the extent possible and realistic for the U.S. marketplace. Technical regulations, including those for legal metrology and conformity assessment procedures (e.g., type evaluation), are covered by these agreements. The NIST Weights and Measures Division provides the primary technical support for the U.S. commercial measurement system.

The decentralized weights and measures system in the United States creates a great challenge to achieve uniformity among the many regulatory jurisdictions. Conflicting regulations, varying interpretations of the same or similar requirements and divergent methods of enforcement seriously interfere with the efficiency of any program and are particularly unfortunate when associated with the administration of a weights and measures law. In the first place such conflicts and disparities throw a great burden upon the manufacturers of weighing and measuring equipment, a burden that is eventually borne by the ultimate consumer through increased costs of the products he buys. In the second place these are most confusing to the business interests of the State, which are forced to conform to whatever requirements may be in force in the locality where a particular transaction takes place. Also, non-uniform requirements confuse the purchasing public, and complicate the enforcement of the law and hamper the officials who are trying to enforce it.

Each state has the authority to establish its own weights and measures requirements. However, significant differences in such a fundamental regulatory function would disrupt commerce in the United States. These differences in laws, regulations, and technical regulations and the subsequent disruptions to commerce contributed to the formation in 1901 of the National Bureau of Standards (NBS), which is now NIST. To maintain the commercial measurement infrastructure and to modify and improve it to keep up with technological advances and the changing marketplace requires major work by and cooperation among industry, NIST, federal, state, and local regulators.

In 1905, NBS hosted the first meeting of weights and measures directors to address these problems. These meetings became formalized as the NCWM. See Section 10.2 The Role of NCWM for more information.

Businesses change over time to become more efficient and competitive, and the needs and interests of consumers also change. Weights and measures officials must be sensitive and responsive to these changes and modify their programs and requirements as a result of changes in the marketplace, in technology, in consumer interests, and the needs of business. Weights and measures officials must continue to pursue their goals of equity in the marketplace and fair competition among companies without their requirements and practices becoming obsolete or an undue burden on industry.

In the first half of the twentieth century, all commercial measuring instruments were mechanical devices. Mechanical devices evolved slowly and had many similarities in design, so the principles of operation were relatively easy to understand. When electronic measuring instruments came to dominate the commercial marketplace in the second half of the century, the evolution of electronic instruments became extremely rapid. Additionally, the software of the instruments changed even more rapidly. Commercial measuring instruments are no longer standalone devices, but contain sophisticated software so the measuring instruments can be integrated into overall business management software. Additionally, the nature of the marketplace has changed.

Early in the history of the United States, local communities were served by local businesses. The ramifications of regulatory actions were similarly localized. However, the marketplace has changed from the nature of direct sale to consumers to the delivery of many products through prepackaged goods. Package labeling and concerns about deceptive packaging became major issues. Many retail stores have grown from "mom and pop" operations into national and international companies. Now, regulatory action in one supermarket may cause reactions through the national chain of stores. Consequently, changes in weights and measures laws, regulations and technical requirements (standards) often have major ramifications on product design, manufacturing and the marketing of consumer products.

The complexity of weights and measures regulation has increased dramatically over the last 50 years, and maintaining compliance with package labeling requirements is an ongoing challenge in a changing and global marketplace. Weights and measures officials must have a wide range of knowledge, and an understanding of the operation of the commercial measurement system, the basic concepts of physics that apply to commercial measuring instruments and the test procedures that they use. The officials must have a basic understanding of statistics used in the procedures for package inspection and must be able to apply statistical analysis concepts to the results of package inspections and device testing. This highly technical work is best accomplished by individuals who are able to dedicate their time to weights and measures duties.

5.0   Standards and Units

The standards and units of measurement that may be used within the United States are specified by Congress. Both the International System of Units (the metric system) and the inch-pound units are permitted for use. Most state weights and measures laws also state that these two measurement systems may be used. At the present time, packaged consumer goods that fall under the U.S. Fair Packaging and Labeling Act (FPLA) are required to be labeled in both inch-pound and metric units. However, the NIST Weights and Measures Division is working with Federal agencies and industry to encourage a change to FPLA to allow packages to be labeled in only metric units to facilitate international trade. Most states already allow packages that fall only under state packaging and labeling requirements to be labeled only in metric units.

States are encouraged to adopt weights and measures laws and regulations that are consistent across the country. The NCWM recommends uniform weights and measures laws for adoption by the states. The technical requirements for measuring instruments are contained in NIST Handbook 44, *Specifications, Tolerances, and Other Technical Requirements for Weighing and Measuring Devices*. The other recommended laws and regulations are published in NIST Handbook 130, *Uniform Laws and Regulations in the Area of Legal Metrology and Engine Fuel Quality*. Both Handbooks 44 and 130 are amended as necessary through the standards development process of the NCWM. Brief descriptions of some of the uniform laws and regulations contained in NIST Handbook 130 are provided below.

6.0   Uniform Laws and Regulations

6.1   Measuring Instruments (Device) Regulation

The measuring instruments (or device) regulation is intended to specify the design and performance requirements of measuring instruments used commercially. (Note that the term "measuring instrument" is synonymous with the term "device" as used in NIST Handbook 44.) The scope of measuring instruments that fall under legal metrology control must be specified. NIST Handbook 44 contains the technical and performance requirements for commercial measuring instruments used in the United States. Each state should adopt the latest edition of Handbook 44. The weights and measures inspector uses Handbook 44 to determine if measuring instruments used in commercial applications comply with the requirements.

Before measuring instruments may be installed in stores or at business locations, most states require that the many types of measuring instruments have type evaluation certificates reporting that the models comply with the requirements of Handbook 44. If a model of measuring instrument has a type evaluation certificate, then the inspectors may focus their inspections on the suitability of each instrument for its application, and verify that it has been properly installed, that the correct operating features are in use, that the measuring instrument is accurate within prescribed tolerances, that the proper software is installed, and that the adjustments are properly sealed or that the audit trail reveals that the metrological features are not being manipulated for fraudulent purposes.

6.2   Type Evaluation Program

The National Type Evaluation Program (NTEP)[5], which is managed by the NCWM, is the type approval authority in the United States. The purpose of a type evaluation program is to verify that measuring instruments have demonstrated compliance with the specifications and performance requirements before they are installed in commercial applications. For many years, type evaluation was not included in the model laws and regulations. By 1967, there were various forms of type evaluation requirements in existence in 14 states, 2 cities, and 1 county. Meeting

---

[5] Additional information about the National Type Evaluation Program can be found at: http://www.ncwm.net.

the needs of so many type evaluation requirements posed additional expenses and trade barriers for manufacturers within the United States. In 1984, NCWM adopted a model regulation known as the Uniform National Type Evaluation Regulation.

Most states require NTEP certificates for measuring instruments for which NTEP conducts evaluations. The NTEP Certificate of Conformance would have no legal value unless states adopt some form of regulation recognizing NTEP.

In the absence of a type evaluation requirement, each weights and measures inspector must conduct a more extensive field inspection whenever a new model measuring instrument is found in use in order to verify compliance with specifications and performance requirements. There are some performance requirements to which compliance cannot be determined in the field, particularly accuracy under a range of environmental influence factors. Consequently, type evaluation is a process through which all measuring instruments are required to meet a minimum set of requirements before they are installed for commercial use. The Uniform Regulation for National Type Evaluation is contained in NIST Handbook 130.

6.3 Weighmaster Law

The adoption of a weighmaster law depends on how the jurisdiction chooses to inspect and verify the accuracy of transactions in which weighing the commodity is a major aspect of the transaction. The objective is to have accurate transactions. One can simply require that the transactions be accurate and then conduct test purchases or reweigh commodities on a sufficiently frequent basis to verify the accuracy of the transactions. It is unlikely that any jurisdictions do enough reweighing of bulk commodities to ensure that the transactions are accurate, so most weights and measures programs have limited assurance that weighing transactions are performed accurately.

The other approach to controlling the accuracy of transactions where weighing the commodity is routine is to have a weighmaster law in effect. A weighmaster law requires that the people performing particular types of weighings be licensed by the state and trained in the proper techniques to perform accurate weighings. The weighmaster is also required to record the important information needed to document each transaction and provide a weight certificate for each weighing. Under this law, the weighmaster is held responsible for the accuracy of each weighing and the weighing process is checked on an infrequent basis to verify that the weighmaster is fulfilling his or her responsibilities. The Weighmaster Law is contained in NIST Handbook 130.

6.4 Method of Sale Regulation

The Method of Sale Regulation specifies the measurement basis on which specific products may be sold. The objective is that similar products should be sold using the same measurement unit so that consumers can make value comparisons among the different sized packages of the same brand of product and different brands of products. For some products, traditional methods of sale within the country may be considered. Additionally, more than one basis of measurement may be permitted (e.g., by weight or count for some produce products) depending upon the traditional methods of sale that exist within the states. However, the objective is that the method of sale shall provide accurate and adequate quantity information that permits the buyer to make price and quantity comparisons. The Uniform Method of Sale Regulation is contained in NIST Handbook 130

6.5 Packaging and Labeling Regulation

A packaging and labeling regulation specifies the information that must be provided to buyers of packaged goods to identify the product and quantities contained in the package, and to facilitate value comparisons. The Food and Drug Administration and the U.S. Department of Agriculture have additional labeling requirements that deal with the ingredient and nutrition labeling of food products, which preempt conflicting state regulations. However, many products fall under the labeling requirements of state regulations only. Therefore, it is important that state packaging and labeling requirements be consistent among the states.

The labeling requirements are contained in NIST Handbook 130, "Uniform Packaging and Labeling Regulations." NIST WMD has issued short publications (NIST SP1020 series) that summarize the packaging and labeling requirements contained in NIST Handbook 130. (See the following link to publications: http://www.nist.gov/pml/wmd/pubs/index.cfm. Packagers can be encouraged to use these documents to help them design packages and labels to contain the appropriate and required information. Weights and measures officials are encouraged to use the documents as aids when inspecting packages and labels for compliance with the requirements.

6.6 Voluntary Unit Pricing Regulation

The Voluntary Unit Pricing Regulation provides guidance to stores on how price and unit of measure information must be presented when posting unit prices. The objective is to provide consumers with cost information in a common measurement unit for similar products so the consumer can evaluate this cost per unit along with other value criteria that the consumer may wish to use. A unit price regulation is preferred over a regulation that specifies package sizes that may be used for particular products.

If packages of similar products are of the same size, then cost comparisons can be made based upon the total price of the package. However, fixed package sizes usually do not meet the needs of consumers, and for the most part have been eliminated for most packaged goods.

A unit pricing regulation has the advantage in that it encourages more products to enter the market to fulfill the needs of different consumer groups, since labeling the unit prices on the shelves of the retail stores gives consumers important information to facilitate value comparisons. However, the burden is then on the operators of retail stores to maintain the unit prices accurately. Additionally, regulatory inspections are often required to ensure that stores are maintaining the accurate posting of unit prices. The Uniform Unit Pricing Regulation is contained in NIST Handbook 130.

6.7 Registration of Service Companies

Service companies that install and repair commercial weighing and measuring instruments should be competent to perform these services. These service companies are typically authorized by the weights and measures regulatory authority to place new and rejected weighing and measuring instruments into commercial service. They are required to send "placed in service" reports to the weights and measures administrator to alert them when new commercial measuring instruments (devices) have been installed or repaired and where they are located so a weights and measures official can conduct an official inspection. Consequently, many states require that these service agencies and their service technicians register with the regulatory authority.

As part of this registration process, the service agency must prove that it has adequate test equipment and field standards to perform the tests and that the service technicians know and understand the legal metrology requirements that apply to the devices that they service so the performance of measuring instruments comply with legal requirements. The Uniform Voluntary Registration Regulation is contained in NIST Handbook 130.

A state needs an adequate service industry network before either a voluntary or mandatory registration regulation can be put in place. This is a particular problem for rural states when towns and cities are far apart and the number of service companies is small. Nevertheless, weights and measures officials are discouraged from attempting repairs or adjustments on measuring instruments because they are trained to perform this task, and due to possible liability and conflict of interest issues. After a registration regulation is adopted, regulatory oversight is required to ensure that registered service agencies are performing adequately. Legal action must be taken to withdraw registrations of service agencies or agents that are not performing tests and repairs adequately.

6.8   Price Verification Program

Most stores use scanning systems at the checkout registers to identify the items being purchased and to look up the prices of the items in a database. It is important that the prices posted on the shelves or marked on the individual items are the same as the prices stored in the computer database, since posted and advertised prices must agree with what the customer is ultimately charged. Under the requirement that prices be accurately stated (see Section 16. Misrepresentation of Price, in NIST Handbook 130), prices posted on a sign or shelf for a product must be the same as the price charged at the checkout stand. When a state operates a price verification program, the compliance rate that is considered minimally acceptable and the sampling procedure to be used to determine compliance should be consistent with the Examination Procedure for Price Verification, which is contained in NIST Handbook 130.

6.9   Open Dating Regulation

An open dating regulation "is to prescribe mandatory uniform date labeling of prepackaged, perishable foods and to prescribe optional uniform date labeling that must be used whenever a packager elects to use date labeling on prepackaged goods that are not perishable. Open dating is intended for use and understanding by distributors, retailers, and consumers when judging food qualities."[6] The Uniform Open Dating Regulation is contained in NIST Handbook 130.

---

[6] NIST Handbook 130, 2011 Edition, *Uniform Laws and Regulations in the Areas of Legal Metrology and Engine Fuel Quality*, "Uniform Open Dating Regulation," page 149.

6.10 Fuel Quality Laws

Many states have fuel quality laws. These laws should be consistent with the uniform law published in NIST Handbook 130. A petroleum quality law and inspection program is even more important than in the past, because of the many alternative fuels and product blends offered on the market. Ensuring that engine fuels and heating fuels are the products that they are stated to be is an important part of weights and measures regulation. This is an example where weights and measures programs have gone beyond the quantitative aspects of transactions to also include quality characteristics of products.

7.0 The Role of the Metrology Laboratory

The accuracy of measuring instruments is an essential component of accurate transactions. Accurate physical standards are required by weights and measures officials and by service companies to test measuring instruments. Over the history of the United States, Congress has periodically authorized and funded grants of new standards and test equipment to the states to serve as the reference standards for the states. The last authorization occurred in 1965. New standards and test equipment were issued to the states, the District of Columbia, and Puerto Rico between 1965 and 1978.

To receive the standards, each state had to provide an adequate laboratory facility and hire a qualified metrologist, who was trained through the NIST WMD Laboratory Metrology Program. Many state metrology laboratories have been accredited through the NIST National Voluntary Laboratory Accreditation Program (NVLAP). As stated in NIST Handbook 143:

> State legal metrology laboratories are custodians at the State level of measurement standards that serve as the basis for ensuring equity in the marketplace and as reference standards for calibration services for indigenous industry. As part of its program to encourage a high degree of technical and professional competence in such activities, the National Institute of Standards and Technology (NIST) Weights and Measures Division (WMD) has developed performance standards and formalized procedures for Recognition of State legal metrology laboratories on a voluntary basis. Certificates of Measurement Traceability are issued upon evaluation of the laboratory's ability to make reliable metrological measurements (principally mass, volume, length, and temperature).[7]

7.1 Physical Standards

The physical standards and the measurements provided by the metrology laboratory are the foundation of the commercial measurement system and the legal metrology regulatory system. Accurate measurement standards are necessary for both service companies and regulatory officials when testing and adjusting commercial measuring instruments in order to provide accurate measurement results. The NIST Weights and Measures Division (WMD) provides laboratory metrology seminars to metrologists who work in state and industry laboratories to ensure correct maintenance and use of standards and equipment and to ensure accurate and

---

[7] NIST Handbook 143, *State Weights and Measures Program Handbook*, 5th Edition (2007), page 1.

traceable measurements. Proficiency tests are regularly conducted among the laboratories to validate the laboratory ability to provide acceptable measurement results. Additional training is provided in meetings of regional metrology groups to keep metrologists current with advances in laboratory metrology and national and international requirements for calibration laboratories so that the laboratories can perform at the acceptable levels. NIST Handbook 44 states:

> **3.3. Accuracy of Standards.** - Prior to the official use of testing apparatus, its accuracy should invariably be verified. Field standards should be calibrated as often as circumstances require. By their nature, metal volumetric field standards are more susceptible to damage in handling than are standards of some other types. A field standard should be calibrated whenever damage is known or suspected to have occurred or significant repairs have been made. In addition, field standards, particularly volumetric standards, should be calibrated with sufficient frequency to affirm their continued accuracy, so that the official may always be in an unassailable position with respect to the accuracy of his testing apparatus. Secondary field standards, such as special fabric testing tapes, should be verified much more frequently than such basic standards as steel tapes or volumetric provers to demonstrate their constancy of value or performance.
>
> Accurate and dependable results cannot be obtained with faulty or inadequate field standards. If either the service person or official is poorly equipped, their results cannot be expected to check consistently. Disagreements can be avoided and the servicing of commercial equipment can be expedited and improved if service persons and officials give equal attention to the adequacy and main-tenance of their testing apparatus.[8]

The field standards used by weights and measures inspectors must be appropriate for the tests to be performed, valid and traceable. Handbook 44 states the following regarding the accuracy of field standards:

> **3.2. Tolerances for Standards.** - Except for work of relatively high precision, it is recommended that the accuracy of standards used in testing commercial weighing and measuring equipment be established and maintained so that the use of corrections is not necessary. When the standard is used without correction, its combined error and uncertainty must be less than one-third of the applicable device tolerance.
>
> Device testing is complicated to some degree when corrections to standards are applied. When using a correction for a standard, the uncertainty associated with the corrected value must be less than one-third of the applicable device tolerance. The reason for this requirement is to give the device being tested as nearly as practicable the full benefit of its own tolerance.[9]

---

[8] NIST Handbook 44, 2011 Edition, Appendix A – Fundamental Considerations, page A-5
[9] NIST Handbook 44, 2011 Edition, Appendix A – Fundamental Considerations, page A-4

In some cases, particularly for volume standards used to test petroleum liquid meters, it is challenging to keep the error and uncertainty of the field standard to less than one-third of the smallest tolerance applied to some refined petroleum liquid meters.

The training provided by the NIST WMD Laboratory Metrology Program covers the appropriate test procedures, control charts, and uncertainty calculations to be used for the calibration and tolerance testing of field standards. Laboratories are required to have implemented management systems, including quality manuals that document all aspects of the laboratory operation, appropriate levels of standards for each type of measurement, and to periodically calibrate the working standards to monitor their stability.

Standards are expected to remain within tolerance during the time that they are in use between calibrations by a recognized or accredited state metrology laboratory. The frequency to recalibrate different standards may vary based upon the nature of the standards and the type and amount of use that the standards receive. However, inspection and calibration of standards should be done regularly to assure the standard is maintained in good working order and in tolerance during the entire calibration interval (cycle). The metrologists and the weights and measures director typically have established time periods for the regular calibration of the different standards used by field inspectors. Recommended baseline calibration intervals are posted on the NIST website at: http://www.nist.gov/pml/wmd/labmetrology/index.cfm in the Newsletter archive resources.

The large cast iron weights commonly used on vehicle scale test units and the 25 lb or 50 lb weights used by field inspectors to test vehicle and platform scales tend to experience more abuse and wear than the stainless steel 1 lb, 2 lb and 5 lb weights used to test scales in delicatessens and checkout stands in supermarkets. Consequently, cast iron weights may require more frequent calibrations (and cleaning and painting) than small weights that are transported in protective carrying cases. It is important to always obtain an "as found" calibration on standards before they are cleaned, painted or have any adjustment made to ensure (and be able to prove) that they have remained in tolerance while in use for enforcement. If standards do not remain in tolerance between calibrations, it is recommended that the calibration interval be adjusted, and additional training be given on the care and handling of standards. The state metrologist will have received training on the calibration intervals for different standards. The following information was taken from the NIST WMD web site for laboratory metrology. (http://www.nist.gov/pml/wmd/labmetrology/index.cfm)

7.2   Calibration Intervals

The NIST website provides information on calibration intervals as follows:

> *Legal requirements.* Each state establishes legal requirements for periodic verification of Class F test weights used for commercial applications. In most cases, this is a fixed interval of one or two years. In some cases, evaluation of historical data has been used to establish other fixed intervals for periodic calibration.

19

Industry/scientific. There is no fixed calibration interval for industrial or scientific applications. For these applications, a calibration interval must be established based on: 1) calibration information, tolerances, uncertainties, and applications at time of test; 2) historical data for weight artifacts showing stability (or lack of stability) with time and use; or 3) use of a measurement assurance program where control standards or check standards are used periodically to verify continued accuracy and traceability or the need for calibration. For example, typical calibration intervals for a 100 gram weight set range from six months to five years (or longer) based on measurement assurance data or historical data from periodic recalibration. For some defense applications, fixed recalibration cycles are established with a one-year period.

Reference standards. There is no fixed interval for recalibration of reference standards in the state legal metrology laboratories. When extensive measurement assurance programs are in place to evaluate the accuracy and traceability of the standards and measurement process during use, the laboratory may evaluate when its standards must be recalibrated based on data available in the laboratory. This data is annually reviewed by the Weights and Measures Division. As a part of each laboratory's measurement assurance program, it is an essential practice to: 1) periodically insert an outside check on the system (such as through a proficiency test); 2) have reference kilograms recalibrated periodically when surveillance programs are in place; or 3) maintain a NIST-traceable control standard that is not used with the same frequency as other working standards or check standards to periodically verify measurement control.

Guidelines. For further assistance in establishing calibration intervals, the National Conference of Standards Laboratories (NCSL) has a Recommended Practice (RP-1) on "Establishment and Adjustment Calibration Intervals" (2010).

Devices. Calibration intervals for balances and scales are typically established in a pattern similar to that for test weights. For example: scales used for commercial applications must be periodically verified as established by state regulations, and balances or scales used for other applications must have verification intervals evaluated based on stability through time and use.

Tolerance Tables. In addition to the specifications (e.g., ASTM E 617, NIST 105-1, OIML R 111-1 [part 1]; OIML R 111-2 [part 2]), which have tolerance tables available with detailed information about choosing appropriate weights and verification periods.

Nevertheless, the field inspectors should transport and handle standards with care appropriate to how the standards are used. Inspectors should be careful not to drop weights, because the weight could cause injury to the inspector or others near the scale under test and may damage the weight so that it may not be within tolerance. Whenever anything happens that may bring the validity of the standard into question, the standard should be submitted to the metrology laboratory for test and certification.

The following suggestions are made regarding the care and use of field standards:

- Field standards shall be handled with care to minimize the potential for damage.

- Field standards, test equipment, and carrying cases shall be kept clean and in good repair to maintain the accuracy of the standards and to present a professional appearance.

- Standards and test equipment should be sheltered from the elements when not in use.

- Cast iron weights should be lifted when moved across a scale platform to reduce the loss of material from the standards that would occur from sliding the weights across the scale platform.

- Cast iron weights should be cleaned and painted on a regular basis (and checked in the metrology laboratory before and after cleaning and painting) to maintain a professional appearance.

- Test measures should be inspected regularly for dents.

8.0   Traceability

Traceability is very important, but the conditions that must be satisfied to establish traceability may be complex. Figure 1 illustrates the chain of traceability, from international standards to the point of consumer purchase.

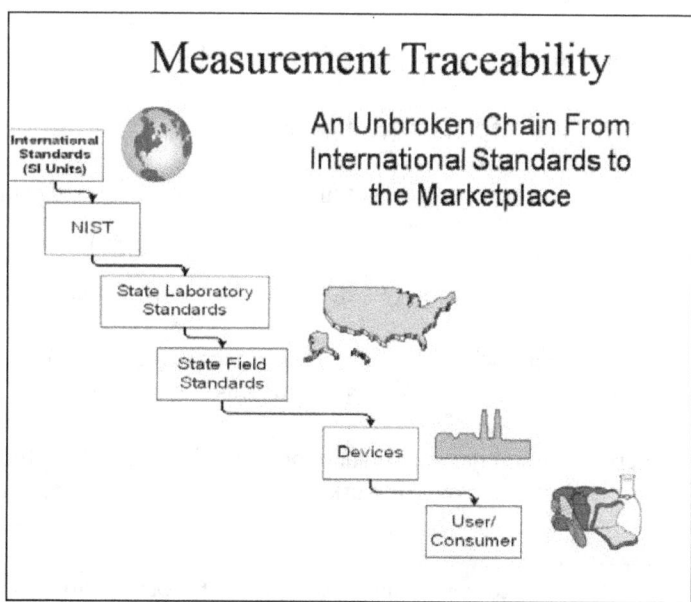

Figure 1. Measurement Traceability

The NIST website provides answers to frequently asked questions (FAQs) about traceability[10]. The following information was taken from the website.

"What is traceability?

The definition of traceability that has achieved global acceptance in the metrology community is contained in the International Vocabulary of Basic and General Terms in Metrology[11] (VIM; 2008):

"...property of a **measurement result** whereby the result can be related to a reference through a documented unbroken chain of **calibrations,** each contributing to the **measurement uncertainty**."

It is important to note that traceability is the property of the result of a measurement, not of an instrument or calibration report or laboratory. It is not achieved by following any one particular procedure or using special equipment. Merely having an instrument calibrated, even by NIST, is not enough to make the measurement result obtained from that instrument traceable to realizations of the appropriate SI unit or other stated references. The measurement system by which values are transferred must be clearly understood and under control.

What do I need to do to support a claim of traceability?

The provider of a measurement result or value of a standard must document the measurement process or system used to establish the claim and provide a description of the chain of comparisons that were used to establish a connection to a particular stated reference.

All valid statements or claims of traceability have the following elements:

- A clearly defined, particular quantity that has been measured;

- A complete description of the measurement system or working standard used to perform the measurement;

- A stated measurement result or value, with a documented uncertainty;

- A complete specification of the stated reference at the time the measurement system or working standard was compared to it;

- An 'internal measurement assurance' program for establishing the status of the measurement system or working standard at all times pertinent to the claim of traceability; and

---

[10] NIST web site URL: http://www.nist.gov/traceability/traceability_toc.cfm, October 2010.
[11] Bureau International des Poids et Mesures, *International Vocabulary of Metrology – Basic and General Concepts and Associated Terms, VIM, 3rd edition, JCGM 200:2008, http://www.bipm.org/en/publications/guides/vim.html.*

- An 'internal measurement assurance' program for establishing the status of the stated reference at the time that the measurement system or working standard was compared to it

An internal measurement assurance program may be quite simple or very complex, the level or rigor to be determined depending on the level of uncertainty at issue and what is needed to demonstrate its credibility. Users of a measurement result are responsible for determining what is adequate to meet their needs. For information and guidance on expressing measurement uncertainty, see http://physics.nist.gov/cuu/index.html"[12]

The NIST policy on traceability provides helpful information regarding the concept of traceability, and can be found at the following link: http://www.nist.gov/traceability/nist_traceability_policy_external.cfm.

Each weights and measures program should incorporate the following into its practices:

- All standards and test equipment that have a significant effect on the accuracy or validity of the inspection results shall be calibrated before being put into service.

- The standards and test equipment shall be tested on a regular basis consistent with the policy of the weights and measures program. Verification periods may be established using the guidelines referenced in the NIST WMD Laboratory Metrology Program.

- Standards requiring cleaning, painting, or repair should be tested before and after these activities are performed to verify continuous in tolerance results for regulatory purposes.

- If standards or test equipment are damaged or the validity of the standards comes into question, the use of the standards and test equipment shall be discontinued until the validity of the standards can be verified by the metrology laboratory.

For further assistance in establishing traceability, consult NIST IR6969, Good Measurement Practice (GMP) 13, Ensuring Traceability (www.nist.gov/labmetrology).

9.0 Recognition and Accreditation of Laboratories

Not every laboratory needs the same measurement capability, because the services that should be provided depend upon the needs of regulatory agencies, the industry, and academia. The cost and benefit of providing specific services must be examined in conjunction with the availability of qualified services from other sources.

In today's environment, laboratories must be able to demonstrate that they produce traceable measurements. In order to do so, laboratories are expected to demonstrate good management, the use of appropriate measurement procedures, properly documented uncertainty statements,

---

[12] NIST web site URL http://www.nist.gov/traceability/suppl_matls_for_nist_policy_rev.cfm, November 2007.

and the use of acceptable management systems through compliance with the international requirements for calibration laboratories through some form of recognition or laboratory accreditation. Consequently, laboratories are expected to satisfy the requirements of ISO/IEC 17025[13] *General Requirements for the Compliance of Testing and Calibration Laboratories.*

The adequacy of the standards system for legal metrology must examine all aspects of the standards measurement process from the laboratory facilities, the environmental controls, the training and proficiency of metrologists, the management system used in the laboratory, the design and traceability of field standards used by regulatory officials and scale and meter service companies, the registration or certification of scale and meter service agents, and the accuracy of commercial measurement devices and the transactions between buyers and sellers.

The detailed requirements tailored to metrology laboratories already exist. A state metrology laboratory that is accredited by the NIST National Voluntary Laboratory Accreditation Program (or by any other International Laboratory Accreditation Cooperation (ILAC) recognized accrediting body), or recognized through the program managed by the NIST Weights and Measures Division is considered to comply with all appropriate and necessary requirements, including traceability, for those measurement services included in their accredited or recognized scope of measurement services and the associated uncertainties. The assessment criteria for the NIST National Voluntary Laboratory Accreditation Program and the program managed by the NIST Weights and Measures Division Laboratory Metrology Program are available on their web sites (NIST HB 150, http://www.nist.gov/pml/nvlap/upload/nist-handbook-150-1.pdf and NIST HB 143, http://www.nist.gov/manuscript-publication-search.cfm?pub_id=904061, respectively).

The state weights and measures director is responsible for maintaining the traceability of the state standards to the national standards.

It is imperative that the traceability of field standards used by the regulatory official is beyond question. The credibility of inspections, test results and regulatory actions depends upon the use of appropriate physical standards to test the measuring instruments and the proper conduct of the inspections and tests themselves.

Metrology laboratories are expected to provide measurement services to other government agencies to support their activities, to service companies that need accurate standards with which to install and repair measuring instruments, to industry for use in research and manufacturing, and to academia to support their research and education activities.

10.0   Roles of Organizations and Officials

10.1   The Role of NIST

The National Institute of Standards and Technology (NIST) is responsible for securing uniformity in weights and measures laws and applications. In order to fulfill this charge, they:

---

[13] ISO/IEC 17025 (2005) "General Requirements for the Competence of Testing and Calibration Laboratories" available at International Organization for Standardization, http://www.iso.org/iso/search.htm?qt=Iso%2Fiec+17025&sort=rel&type=simple&published=on

- Serve as technical advisors to technical committees and national working groups

- Publish model laws, regulations, procedures and other documents for use in weights and measures enforcement

- Provide technical advice and support to regulators and industry

- Conduct training for metrologists, regulators and industry

- Represent U.S. interests in international legal metrology organizations.

NIST also provides calibration services to the states to ensure traceability in commerce.

10.2   The Role of NCWM

The NCWM is an organization of state and local weights and measures officials, and representatives of industry, consumers, and federal agencies that work together to develop uniform weights and measures laws and regulations, which are published as NIST Handbooks 44, 130 and 133.

The technical committees of the NCWM address weights and measures laws, regulations, device requirements, weights and measures administration, training, and enforcement policies and procedures to promote fair competition and consumer protection regarding transactions based on weighing or measurement. Four regional weights and measures associations[14] support the NCWM. The regional associations develop issues and facilitate participation by local industry and weights and measures officials before the issues are submitted for review and action at the national level by the NCWM. (See the Introduction section in NIST Handbook 44 or 130 for more information regarding the process for revising the Handbooks.)

10.3   The Role of the Fuel Quality Laboratory

Many weights and measures programs are responsible for fuel quality. For some states, this means managing and operating a fuel quality testing laboratory. For others, it means outsourcing the testing of fuel samples that are collected by weights and measures officials. In either case, the program must have procedures in place for the collection and safe handling of fuel samples.

---

[14] The regional weights and measures associations consist of weights and measures officials representing their states. The four regions are: Central Weights and Measures Association (CWMA) representing Illinois, Indiana, Iowa, Kansas, Michigan, Minnesota, Missouri, Nebraska, North Dakota, Ohio, South Dakota, and Wisconsin; Northeastern Weights and Measures Association (NEWMA) representing Connecticut, Maine, Massachusetts, New Hampshire, New Jersey, New York, Pennsylvania, Rhode Island, and Vermont; Southern Weights and Measures Association (SWMA) representing Alabama, Arkansas, Delaware, District of Columbia, Florida, Georgia, Kentucky, Louisiana, Maryland, Mississippi, North Carolina, Oklahoma, South Carolina; Tennessee, Texas, U.S. Virgin Islands, Virginia, and West Virginia; and Western Weights and Measures Association (WWMA) representing Alaska, Arizona, California, Colorado, Hawaii, Idaho, Montana, Nevada, New Mexico, Oregon, Utah, Washington, and Wyoming.

New laws regulating engine fuel quality have increased the need for a robust fuel quality program and many states have increased funding for the monitoring of fuel quality. This is primarily due to the introduction of biofuels into the marketplace, though monitoring other characteristics of fuel is also of importance. Because some states have mandated the inclusion of minimum percentages of biodiesel or ethanol in engine fuels, some weights and measures programs have been tasked with enforcing these mandates through sampling and testing of product.

In the field, many officials test for water contamination of fuel storage tanks on site, and may also have portable octane testing equipment for screening purposes. Special sampling containers are usually provided to officials to facilitate the collection of fuel samples for analysis in the laboratory.

A fuel quality laboratory must have appropriate equipment and specially trained personnel to perform the myriad of tests on various engine fuels. Examples of tests include flash point, octane, cetane, cloud point, distillation, and checking for contaminants. Whether these tests are performed by the state or outsourced, the weights and measures program must have the capability to evaluate test results in order to enforce the law.

10.4 The Role of Manufacturers of Measuring Instruments

The manufacturers of measuring instruments, by necessity, have had a close working relationship with the weights and measures regulatory authorities. Measuring instruments have traditionally been the center of attention of weights and measures officials, because the measuring instruments are the basis for the determination of quantity in commercial transactions.

Weights and measures officials typically focus on measuring instruments used in direct sales to consumers, such as gasoline and diesel fuel dispensers in service stations, meters on trucks that deliver fuel to homes and businesses, point-of-sale scales used at the checkout stands of supermarkets, computing scales used in delicatessens and farmers markets, and vehicle scales used to weigh such products as grain, sand, and gravel.

Scales and meters used to prepackage consumer goods are usually not tested by weights and measures officials, because the packaged goods along with their net content declarations are the basis for the commercial transactions. However, the manufacturers of measuring instruments typically make both the commercial measuring instruments used in direct sales and the noncommercial ones used in packaging consumer goods. The accuracy of scales and other measuring devices used in packaging is critical to ensuring that the packer is employing good manufacturing practice. The U.S. Department of Agriculture's Food Safety and Inspection Service requires all scales used in packaging plants to meet the requirements of NIST Handbook 44.

The Recommendations (international standards) of the International Organization of Legal Metrology (OIML) and NIST Handbook 44 provide extensive design and performance specifications for measuring instruments. Changes in specifications have a major and immediate impact on the manufacturers of commercial measuring instruments. The extensive use of software in measuring instruments complicates the process of modifying measuring instruments to meet new or revised requirements because of the extensive software testing that must be performed in addition to any hardware modifications that may be required.

The manufacturers strive to respond to the needs and requests from their customers (i.e., the users of measuring instruments) and still comply with weights and measures legal requirements. The device manufacturers often represent the interests of their customers when they participate in the activities of the National Conference on Weights and Measures and OIML, so that the appropriate balance of regulation, advances in technology and the needs of users of measuring instruments are satisfied in the most appropriate manner.

10.5   The Role of Consumer Product Manufacturers

The growth in the use of packaged consumer goods combined with national and international distribution makes many more consumer products available to more people than ever before. However, consumer product manufacturers (who will be referred to as packagers) have the responsibility of complying with myriad packaging and labeling requirements in each country in which they market. The labeling requirements routinely include weights and measures requirements, (i.e. Product Identity, Net Quantity of Contents, and Declaration of Responsibility), but also requirements for ingredient and nutritional labeling, when applicable, and the appropriate warning labels and instructions for use as needed. Packagers often have to label packages to satisfy conflicting requirements from different countries and in the language of different countries.

Below is a typical list of labeling issues and differing requirements that are routinely addressed by multinational consumer product packaging companies as reported in a presentation given to legal metrology officials representing the Americas.[15]

Observations on marketing products in the Americas:

- Consumer product regulations differ

- Country labeling requirements differ

- Differing label requirements increase time to market

- Differing label requirements increase cost to market (or prevent the product from going to market)

- Differing label requirements increase label clutter

- Overly specific label content/format requirements create barriers

---

[15] Presentation by Chris Guay, Proctor and Gamble, Symposium on Legal Metrology in the Americas, 2003

Net content statements may also differ in the following respects:

- Units

- Language

- Font size

- Number of significant digits

- Punctuation

Also of note:

- Most countries in Americas require metric but permit supplementary inch-pound units optionally.

- Some Caribbean islands require inch-pound units.

- The United States requires inch-pound and metric.

- Inch-pound volume units (gallon, quart, pint, fluid ounce) are not equivalent between Canada and the United States.

It is obvious from the list above that net content and packaging and labeling requirements have ramifications far beyond national borders. Weights and measures officials must consider the ramifications associated with new or revised requirements for prepackaged goods. Differing or conflicting requirements have a cost associated with them. Packagers may choose not to market some products in some countries because the potential market for the product may not justify the cost. Unfortunately, this has the consequence of denying consumers access to some products and reduces choice and competition, which are negative effects for consumers.

In the United States, packagers must satisfy federal and state labeling requirements, depending upon which products fall under the authority of a particular agency. Furthermore, packages must meet the average net weight requirement and the Maximum Allowable Variations requirements for individual packages. Packagers must demonstrate that they use good manufacturing and distribution practices to comply with both federal and state packaging laws.

While federal and state laws for net content requirements are essentially consistent, the majority of enforcement of net weight requirements for prepackaged goods is performed by the state and local weights and measures inspector. Weights and measures officials frequently conduct net content inspections in retail stores, where the packages available for inspection may represent only a tiny fraction of the production lot of the packager. Checking the net contents of packaged goods is another complex area of weights and measures enforcement. NIST Handbook 133 (2011) states that:

"Testing packages at retail outlets evaluates the soundness of the manufacturing, distributing, and retailing processes of the widest variety of goods at a single location. It is an easily accessible, practical means for State, county and city jurisdictions to monitor packaging procedures and to detect present or potential problems. Generally, retail package testing is not conducive to checking large quantities of individual products of any single production lot. Therefore, follow-up inspections of a particular brand or lot code number at a number of retail and wholesale outlets, and ultimately at the point-of-pack are extremely important aspects in any package-checking scheme. After the evaluation of an inspection lot is completed, the jurisdiction should consider what, if any, further investigation or follow-up is warranted. At the point-of-sale, a large number of processes may affect the quality or quantity of the product. Therefore, there may be many reasons for any inspection lot being out of compliance. A shortage in weight or measure may result from mishandling the product in the store, or the retailer's failure to rotate stock. Shortages may also be caused through mishandling by a distributor, or failure of some part of the packaging process. Shortages may also be caused by moisture loss (desiccation) if the product is packaged in permeable media. Therefore, being able to determine the cause of an error in order to correct defects is more difficult when retail testing is used."[16]

10.6   The Role of Service Companies

The companies that sell, install and maintain commercial measuring instruments are essential to the infrastructure of the commercial measurement system, because these companies provide the service to maintain accurate measuring instruments that comply with all weights and measures (specifications and use) requirements. Measuring instruments that meet all accuracy, specifications and use requirements are called "accurate and correct." A strong weights and measures regulatory program ensures fair competition among businesses and enforcement ensures that owners of commercial measuring instruments get the routine service and repair needed to maintain measuring instruments in accurate and correct condition.

A significant number of service companies are needed within a weights and measures regulatory jurisdiction to create sufficient competition among companies, with the desired benefit of good value for the cost of the service. Companies that service measuring instruments face the same challenges as other types of service companies; that is, finding skilled and competent service technicians at reasonable salaries. The service technicians must not only be knowledgeable in the mechanics, electronics and software of measuring instruments, but they must be skilled and efficient in the troubleshooting and repair of these instruments.

Furthermore, the service technicians and the owner of the service company must be knowledgeable in the weights and measures requirements and test procedures that apply to commercial measuring instruments. A number of scale service companies have been accredited to ISO Standard 17025. This means that these service companies have been evaluated to the criteria of ISO 17025 and meet the requirements, including those requirements to have and use documented test procedures, have adequate field standards, ensure adequate training of their

---

[16] NIST Handbook 133, *Checking the Net Contents of Prepackaged Goods*, 2011, page 1

technicians, and have and use an acceptable quality/management system in their service operation. However, being an accredited service company does not guarantee knowledge of weights and measures requirements.

The service companies must:

- Sell measuring instruments that are appropriate for use in particular applications;

- Sell only measuring instruments that have valid type approval certificates for commercial applications;

- Properly install the measuring instruments; and

- When necessary, train the owners and operators of the measuring instruments in the proper use of the device to ensure accurate transactions.

Service companies are often responsible for placing measuring instruments into service after a weights and measures official has rejected the device for inaccuracy or failing to meet other weights and measures requirements. Consequently, the service company technicians must be sure that measuring instruments are accurate and correct after installation, service, or repair. Service companies typically test and service commercial measuring instruments much more frequently than weights and measures officials conduct enforcement tests on measuring instruments. Therefore, the service companies fulfill a critically important role in maintaining the accuracy of measuring instruments used in the commercial measurement system. Thus, a close working relationship among retailers, service companies and regulatory officials is beneficial to all parties involved.

The companies that sell and service measuring instruments must ensure that the appropriate measuring instruments are sold and placed into service in the correct applications. Some service companies may be authorized to sell measuring instruments made by specific manufacturers. The measuring instruments that they sell should have Certificates of Conformance from the National Type Evaluation Program that is managed by the National Conference on Weights and Measures. Furthermore, the service company must make sure that each measuring instrument is suitable for the application for which it is to be used. Upon installation, the service company must select the appropriate metrological features and parameters for each application. Service companies are frequently allowed to place new and repaired measuring instruments into service (commercial use) prior to an "official" test conducted by a weights and measures official. In these cases, the service company may be required to file a "placing in service" report to the state.

Requirements of a service company:

- Traceable standards

- Trained technicians

- Periodic calibrations

Requirements of a registered service program include:

- Periodic training/exam

- Annual registration/fee

- Adequate oversight

- Authority to revoke permits

The managers of weights and measures programs should know which service companies provide good service and which ones do not. When a state has an adequate oversight program of service companies, the weights and measures program should take action against service companies that provide poor service, that do not install devices suitable for each application or that do not properly install new measuring instruments. If a state has a registration program for service companies, then the registration of poor service companies should be suspended or rescinded for cause. If corrective action is not taken against poor performers, it undermines the ability of good service companies to compete and ultimately hurts the owners of measuring instruments because they are not getting the quality of service for which they pay.

Unfortunately, one or more service companies within a state may not fulfill their responsibilities at an acceptable level. If a state has a service company registration program, but the state has never taken corrective action against a poor service company, then one may question whether or not the state registration program is meaningful.

A registration program for service companies should clearly specify the certification requirements and those requirements must be uniformly applied to all service companies. The service technicians must know the appropriate test procedures and conduct adequate tests to find any deficiencies in the performance of the measuring instruments and then perform the correct repairs to rectify performance problems. Additionally, the service technician must be able to identify and correct any areas of noncompliance with the Handbook 44 specifications that apply to the measuring instruments. Consequently, the service technicians must know the relevant weights and measures requirements that apply to each type of measuring instrument and application of the measuring instruments.

10.7   The Role of Weights and Measures Officials

The primary function of the weights and measures official is as an enforcement authority. The owners of commercial measuring instruments are responsible for maintaining the accuracy of the devices. Weights and measures officials are not service representatives and should not attempt to adjust or repair measuring instruments, due to liability and potential conflict of interest. For example, if an official adjusted and approved a device and the device owner later finds the device is giving away product, the device owner may seek damages from the weights and measures program. Weights and measures officials are expected to be experts in the uniform application of weights and measures laws, regulations and technical requirements, the proper test procedures and the weights and measures requirements applicable to measuring instruments, packaging and labeling requirements, methods of sale of commodities, and all other legal metrology requirements that are applied in their jurisdiction.

Weights and measures officials in each jurisdiction must cooperate with other weights and measures officials in other jurisdictions to coordinate inspection activities whenever problems are found, so that enforcement actions will be more effective. Companies routinely manufacture, distribute, and sell in many states and often across the country. Weights and measures officials must strive to apply weights and measurement requirements correctly, consistently and uniformly across the country (and to the extent possible, internationally) so the commercial measurement system will work efficiently and effectively.

Weights and measures programs usually try to gain compliance with requirements using the lowest level of effort and regulatory action. The most efficient approach to gain compliance is when each business owner is careful and conscientious to use good measurement procedures and practices. Consequently, weights and measures officials will frequently "educate" business owners and device operators on the weights and measures laws, regulations and the responsibilities of the business owner and device operator to properly use the measuring instrument to produce accurate transactions. Some weights and measures programs have formal outreach programs to explain these subjects to the upper management of corporations or store chains in an effort to achieve compliance through the efforts of the businesses themselves.

Weights and measures programs have a variety of regulatory options available for enforcement. These are generally applied in an escalating manner over time if initial efforts are not successful to gain compliance. The lowest level of enforcement for relatively minor violations may be a warning issued to the business for a violation along with regulatory action to correct the problem. In the case of an inaccurate or incorrect measuring device, this may involve rejecting the measuring instrument for repair within a specific amount of time as established by the inspector or program policies. Some jurisdictions may remove the measuring instrument from service until the repair or adjustment is made. In the case of serious infractions, the inspector may seize the measuring instrument. In the case of prepackaged goods, the inspector may order the packages off sale until the packages are reweighed and correct content declarations applied or the packages are returned to the packager.

A higher level of regulatory action may be the imposition of fines or penalties. Many jurisdictions have the authority to issue civil penalties that vary depending upon the seriousness of the violation. The highest level of penalty is usually criminal prosecution of the business owner with all options available to the court to determine the appropriate consequences for the violations.

> **Rejection and Condemnation of Commercial Weighing and Measuring Devices.** When a weights and measures official finds, as a result of his inspection and test of a commercial device, that it cannot be approved for use, he has two options. He can "reject" or "condemn for repairs," or he can condemn outright; his selection of which option to pursue is governed by the character of the conditions found.

Rejection and Rejection Tags. If, in his best judgment, the official believes that the device in question can be repaired and put into proper condition for use he temporarily puts it out of use—until repairs have been made and the device has been retested and approved. This is referred to as "rejecting" or "condemning for repairs." When equipment is so rejected it is suitably marked by the official to indicate this fact—unless the repairs are to be begun immediately. The customary mark is a tag (occasionally an adhesive label) of distinctive color, usually red, setting forth (1) the fact of rejection, (2) the reasons for rejection, (3) the penalty for commercial use before repairs have been made and the device has been reexamined and sealed, and (4) the time limit set for the making of repairs. The tag is signed by the official and is attached to the rejected device in a prominent position by means of the lead-and-wire seal, but not in such a way as to interfere with the making of the necessary repairs. The operator is then fully advised as to the situation and given all necessary instructions.

Follow-up on Rejected Equipment. The time to be allowed for making repairs will differ with circumstances. In a city where service men are available at all times, 5 or 10 days is usually an ample period; in a country district 30 to 60 days may not be unreasonable. In the fixing of this period the official should be given discretionary powers, and he should be careful to allow ample time for the work to be done. In order to follow up cases of rejected equipment, the rejection tag is sometimes made with a perforated stub, which can be filled out with the name of the operator, a description of the device, and the date when repairs should be completed, this stub being retained by the official for his follow-up record; or this same result may be accomplished by retaining a copy of a special "rejection report" when this is used. Needless to say, the official should check up on rejected equipment shortly after the date when repairs should have been completed, and if there is evidence of improper use, or if the operator is negligent about having the repairs made, the official should take whatever action is best suited to the circumstances.

Retests and Permits for Use. In the case of the state inspectors covering large territories it is usually impracticable, on account of the expense involved to follow up matters of this kind as promptly or effectively as can be done in the city or other small territory. Sometimes a service representative or repair man, who may be registered with or licensed by the department, may make the necessary repairs and place the device into service for a period of time before an official test is conducted.

Discarded Rejected Equipment. It frequently happens that when a device is rejected the owner prefers to buy new equipment rather than to have the old equipment repaired. In such cases the rejected device is often turned in as part payment on the new equipment and so passes into the hands of a dealer in weighing or measuring devices. When this occurs the interest of the weights and measures official in the equipment in question does not cease; he should be just as careful in seeing that proper repairs are made before the device is again placed in commercial use as though it had remained in the hands of the original owner, and he should exercise strict control over all reconditioned equipment handled in his territory.

Condemnation of Equipment. As to outright condemnation, this action is taken with relation to equipment that, because of mechanical deterioration or construction deficiencies, is in such bad condition that in the best judgment of the official it is impractical so to recondition it that it will meet specification and performance requirements. When a device is condemned, the official frequently confiscates and destroys it.

Authority to "seize and destroy" is customarily granted to the official by his law with respect to equipment that he condemns and also with respect to equipment that he has rejected but that the owner has not had properly repaired within the specified time limit. This authority should be exercised by the official with discretion. He should keep in mind the property rights of an equipment owner, and cooperate in working out suitable arrangements whenever it is thought practicable for an owner to realize at least something from equipment that has been condemned. In cases of doubt the official should initially reject rather than condemn.

As in the case of equipment approved for use, the official should keep complete records of all equipment rejected or condemned, the reasons for the action taken, and the ultimate disposition of the equipment. As mentioned earlier, follow-up records are also essential in the case of "rejected" apparatus

It is important to emphasize the complexity of the responsibilities of the weights and measures officials and weights and measures programs overall. The weights and measures programs play a critical role in the development and maintenance of the infrastructure of the commercial measurement system. Weights and measures directors, supervisors and individual weights and measures officials should participate in the meetings and discussions of regional weights and measures associations and those of the NCWM. These meetings provide opportunities for officials to get exposed to views and problems experienced by others, and allow participants to be active in developing or revising laws and regulations. The meetings also help provide a broader perspective of weights and measures issues than is possible through exposure to only one jurisdiction.

Communication and awareness of the problems and actions taken in other jurisdictions promote uniformity among weights and measures officials, and each of the stakeholders can provide useful input. Industry provides vital technical and marketing expertise into the discussion of weights and measures issues. Industry also provides valuable input on the ramifications and costs associated with changes to weights and measures requirements. Retailers provide important information on changes in the marketplace and the impact of proposed changes of weights and measures requirements on their businesses. Consumers contribute valuable information regarding their priorities on particular issues and the information that they find necessary to make value comparisons in the marketplace.

11.0 Location of Weights and Measures within an Organization

A specific location for weights and measures programs within state government agencies is not recommended by NIST or the NCWM. The important aspect of organization is that the weights and measures program be located in an agency with similar goals and responsibilities so that the regulatory efforts to ensure a fair and competitive marketplace do not conflict with other program responsibilities of the agency. Additionally, the value placed upon the weights and measures program should be equal to the value of other responsibilities of the agency.

One common problem that several weights and measures programs have reported is that the weights and measures program is incorporated in an agency whose focus is on the promotion of state resources. Sometimes the regulatory nature of weights and measures is considered counterproductive to marketing and promotion. This view of weights and measures is patently incorrect and damaging to the effective conduct of weights and measures programs. Active and strong weights and measures programs ensure that all businesses must comply with the same requirements, establish a "level playing field" for competition, establish credibility to the marketplace, provide the incentive for businesses to maintain accurate measuring instruments, and provide consumer protection against both intentional and unintentional errors in commercial transactions. A healthy weights and measures infrastructure is essential for a healthy commercial measurement system.

12.0 Program Scope

So many areas of the economy are affected by weights and measures that the program itself needs to be pervasive. The importance to a community of adequate weights and measures supervision cannot be overestimated. Next to the personal safety and health of the people, one of the most important of the fundamental obligations of the state or municipality to its citizens is the regulation of commercial weighing and measuring instruments and the exercise of a reasonable control over the users thereof.

The commercial measurement system is huge, and the market segments that should be inspected by weights and measures officials are almost unlimited. Weights and measures directors must determine which areas are going to be inspected and where the inspection resources can best be utilized. When businesses adopt a culture of honesty and make the commitment to comply with weights and measures laws and regulations, the work of the weights and measures regulatory official becomes much easier. It is believed that a much higher compliance rate can be achieved through voluntary compliance than through enforcement actions to force the businesses to comply. Fortunately, most businesses operate ethically with the goal of complying with all applicable laws and regulations.

Weights and measures officials must remember that they serve both businesses and consumers. The goals of providing consumer protection, ensuring fair competition among businesses and facilitating interstate commerce and international trade require that the weights and measures program maintain a balance of interests of industry, consumers and officials. While weights and measures is fundamentally a regulatory activity, weights and measures officials are encouraged to take the time to educate business owners and users of measuring instruments in the weights and measures requirements and their responsibilities to comply with the requirements.

Some weights and measures officials consider their primary objective to be consumer protection, because consumers have limited ways to check the accuracy of transactions. In fact, the weights and measures official is an unbiased third-party that oversees the commercial marketplace to ensure equity in transactions for both the buyer and seller. The goal is to have accurate transactions that do not favor either the buyer or the seller

Even though only a small number of merchants will engage in fraudulent practices, or unintentionally commit errors in measurement, such actions can create a serious problem in the community. The role of weights and measures officials is to protect the interests of both buyers and sellers. Because commercial transactions are fundamental to the necessities of life for citizens, the state has a responsibility to stop unfair practices and pursue legal action if necessary. The weights and measures official stands between the buyer and seller to see that the interests of both are safeguarded. He or she is the impartial arbiter who may be called upon by either party to establish the actual amount of merchandise or service in question, to determine the condition of the weighing or measuring instruments involved, or to take suitable steps to stop an unfair practice or bring about the legal punishment of an offender.

The weights and measures program should be broad based and comprehensive, in that all segments of the commercial measurement system are addressed. For example, programs that focus only on the inspection and test of measuring instruments will ignore the huge segment of packaged goods that represent a major economic part of the commercial measurement system. Similarly, method of sale, unit pricing, and packaging and labeling requirements must be enforced so that consumers are provided with the information needed to make value comparisons. If funding is inadequate to address all segments, the following information should be available and considered when establishing the scope of the weights and measures program:

- Priority device inspection areas are known and data used to justify work priorities;

- Economic impact of the devices; and

- Experience and inspection records regarding levels of noncompliance;

Weights and measures inspections are targeted to:

- Focus on the biggest, highest priority problems;

- Focus on the sectors of the commercial measurement system with poor compliance rates and histories;

- Focus on commercial sectors of greatest economic importance;

- Include the broad range of commercial measurements falling under weights and measures authority without neglecting important commercial activities;

- Verify transaction accuracy as well as device accuracy;

- Include regular package inspection activities;

- Focus on products with high rates of noncompliance, although all types and items are tested periodically to ensure adequate sampling of the marketplace;

- Include price verification inspections; and

- Have the flexibility to shift inspection resources from sectors of high compliance to sectors with limited inspection activities.

The major economic activities in a geographic area affect how weights and measures resources are allocated. Typically, large grain producing states will dedicate significant resources to test the accuracy of vehicle scales used to weigh truckloads of grain and these states will likely have grain moisture meter and grain protein analyzer test programs. States generally assign significant resources toward the inspection of retail stores and supermarkets in metropolitan areas because towns and cities are major economic centers for the surrounding region. However, businesses in sparsely populated areas also need regular inspections, since all businesses and consumers need and benefit from weights and measures inspections. Another valuable source of economic information for each state is the Economic Census conducted by the U.S. Census Bureau. The economic census provides detail and insight into the different and changing components of the economies of each state.

A list of typical inspection areas (disciplines) is shown below in Table 2, but the list is not all-inclusive. Enforcement activities should be fair and equally applied to all companies in the same segment of the commercial measurement system. The comprehensiveness of a weights and measures program can be evaluated by determining to what extent the program conducts inspections in each discipline that is a significant part of the state economy. In many instances, the accuracy of quality measurements, such as those made by grain moisture meters and carcass evaluation instruments have a greater impact on the price paid for the commodities than the accuracy of the scales used to weigh these commodities.

Table 2. Measurement Activities and Instruments

Activity	Measuring Instruments

■ Net content of packages
  ➤ Standard pack packages
  ➤ Random weight packages
  ➤ Sale by volume
  ➤ Sale by area
  ➤ Sale by length
  ➤ Sale by count
  ➤ Inspection of commodity purchases by state institutions
 	
■ Price verification
■ Undercover test purchases
■ Fuel quality inspection
■ E-commerce methods of sale and accuracy of delivered products
■ Packaging and labeling
■ Method of sale of commodities
■ Device inspection
	■ Weighing devices
	  ➤ Retail computing scales
	  ➤ Point-of-sale scales
	  ➤ Shipping scales
	  ➤ Platform scales
	  ➤ Vehicle scales
	  ➤ Railway track scales
	  ➤ Hopper scales
	  ➤ Precision scales (Class I and II)
	  ➤ Belt-conveyor scales
	  ➤ Automatic weighing systems
	■ Measuring devices
	  ➤ Retail motor-fuel dispensers
	  ➤ Vehicle-tank meters
	  ➤ Loading-rack meters
	  ➤ Liquefied petroleum gas (LPG) meters
	  ➤ Water meters (or water sub-metering meters)
	  ➤ Hydrocarbon vapor-measuring devices
	  ➤ Cryogenic liquid meters
	  ➤ Milk meters
	  ➤ Mass flow meters
	  ➤ Carbon dioxide liquid meters
	■ Other devices
	  ➤ Taximeters
	  ➤ Grain moisture meters
	  ➤ Grain protein analyzers
	  ➤ Carcass evaluation instruments in processing plants
	  ➤ Multiple dimension measuring devices (shipping industry)

13.0   Program Management

The managers of the weights and measures program, that is, the weights and measures director, the program managers and/or supervisors, are responsible for the proper conduct of inspections and testing, and the allocation of resources. Obviously, the program managers must operate within the management directives of the agency of which the program is a part and the program is constrained by the resources allocated to it by the state legislature and agency administrator. Therefore, legislators and agency administrators need to understand and appreciate the scope and importance of the weights and measures program to ensure fair competition, to provide consumer protection, and to facilitate interstate commerce and international trade.

Weights and measures laws, regulations, and enforcement should be viewed from both a national and international perspective. While most weights and measures regulatory authority rests with the individual states, differences in weights and measures requirements, inspection methods, test procedures, and interpretations of the same requirements may cause problems for interstate commerce and international trade. Weights and measures enforcement is much more effective if it is unified nationally rather than consisting of many independent activities. Furthermore, weights and measures regulatory programs are encouraged to cooperate with both neighboring states and across the country to share information and increase the effectiveness of regulatory activities.

National unity in weights and measures can be achieved even with the regulatory authority resting with the individual states. However, there must be a commitment by each state program to national uniformity. Furthermore, the development of technical regulations and the resolution of weights and measures issues must be addressed from a national perspective and not just from the perspective of the state or local jurisdictions. At times, the commitment to national uniformity requires that local interests become secondary to national interests. Regional associations and the NCWM serve as forums in which local, state, and regional concerns are shared. Such interactions help to determine whether an issue is of national interest and, therefore, merit an amendment to the standards.

The director must ensure that inspectors receive adequate supervision and training to achieve uniformity in inspections and enforcement policies. The director must communicate and coordinate with other weights and measures directors to follow up on areas of noncompliance found in the marketplace. Keeping up with changing technology and passing this knowledge on to the field inspectors is another challenge.

For a number of jurisdictions, the weights and measures office is the link through which the field inspectors determine whether or not measuring instruments installed in the field have NTEP Certificates of Conformance (COC). Ideally, inspectors access the COCs on the internet using laptops in the field. The weights and measures program must have inspection report forms (either hard copy or electronic if computers are used in the field) to capture important information that will allow the director to determine the levels of compliance in different inspection activities and identify any trends in compliance that may be developing.

Supervisors should work with each employee on a regular basis or have other effective ways to ensure that the proper inspection and test procedures are being used, that enforcement actions are taken consistent with established policies and to see that the inspectors are conducting their inspections in an efficient manner. The supervisors must be sure that the field inspectors are presenting the desired image that the program wishes to portray. Good people skills are required in inspectors along with common sense.

14.0   Administrative Functions

14.1   Budget

Numerous administrative functions are or should be part of every weights and measures program. Since obtaining adequate funding is fundamental to running any program, the state director must develop effective legislative budget requests based on sound data that:

- Demonstrate the scope, effectiveness and benefits of the program;

- Include clear and measurable program objectives;

- Identify the inputs for different activities, the expected outcomes from these efforts, and be able to demonstrate the results actually obtained as a result of these efforts; and

- Include statements of support from stakeholders for these activities or initiatives.

The weights and measures director must know the legislative and agency priorities of current administrations and use this knowledge to tailor and justify the budget request to show how the weights and measures program contributes to achieving the objectives. The director may be competing with other agencies of the state government to get funding. Effective and persuasive budget proposals increase the chances for success, but do not guarantee success. The weights and measures directors that have successfully communicated to the legislature the importance of weights and measures activities for the economic benefit of the state, industry and consumers can point to a healthy budget as one measure of management success.

14.2   Data Management

Weights and measures directors must collect and interpret a large amount of information, including:

- Retail business data, to include location, contacts, etc.

- Commercial device data, to include location, manufacturer, model, etc.;

- Assignment and allocation of staff throughout the state to most effectively conduct inspections;

- Inventory of equipment used to perform inspections, not only of measuring instruments, but also for package checking, method of sale inspections, price verification inspections, packaging and labeling inspections, complaint investigations, undercover investigations, etc.

- Inspection data;

- Service company data;

- Laboratory data; and

- Complaint data

An efficient and effective computerized data management system is needed to capture the inspection results, service company information and the logging of complaints. It is essential that there be a way to analyze this summary data. Simply capturing this information without using it is a waste of resources and valuable information.

Similarly, the program must have a laboratory data management system capable of maintaining and retaining laboratory test results and records (for both metrology and fuel quality laboratories). Ultimately, ideal data management systems will allow for streamlined annual reporting to policy makers and the legislature on the performance of the weights and measures program.

14.3   Uniform Test Procedures

A program must have uniform and well-documented test procedures for package inspection and the inspection and test of measuring instruments. The test procedures must be consistent with NIST Handbooks 44, 130 and 133, appropriate, technically sound, and have a valid basis in statistics when sampling plans are used. Valid statistical techniques must be used when analyzing inspection results. Statistical assessments should be made when determining the appropriate enforcement action to be taken based upon the number and types of errors that are found in inspections.

The need for uniform procedures was recognized many years ago, so recommended inspection and test procedures have been published in different NIST handbooks. NIST Handbook 133 provides detailed procedures to determine net content of prepackaged products sold by weight, volume, area, count and length. NIST Handbook 112 provides examination procedure outlines for measuring instruments. These procedures have been developed with the participation of the manufacturers of the measuring instruments used commercially.

NIST Handbook 130 has a recommended procedure for conducting price verification inspections. Weights and measures officials that use these recognized procedures have a significant level of confidence in their validity. If weights and measures officials (i.e., directors and program managers) develop their own procedures for inspection and test, then the officials are responsible for doing their own research and consultation with industry to demonstrate the technical and statistical validity of their inspection and test procedures.

Similarly, enforcement policies must be documented to promote consistency among the actions taken by weights and measures inspectors. Inspectors must still use their judgment to determine the appropriate enforcement action in a particular situation, but any deviation from state enforcement policy should be coordinated with the official's supervisor.

14.4   Communication Programs

Several weights and measures directors have reported that a key factor to their success has been regular and timely communication with the businesses that are regulated by the program. The benefits are that the business owners know about the weights and measures laws and regulations, know what to expect in inspections, are informed and consulted about changes in the program so they are not surprised by changes, understand their responsibilities to comply with the requirements, and understand the goals of the programs that affect their businesses.

Business owners appreciate the openness of the weights and measures program and the opportunity to provide input to the weights and measures directors. These businesses are critical to generating support for weights and measures budget requests when these are submitted to the legislature. They also recognize the benefits of a strong weights and measures inspection program to maintain fair competition in the marketplace and to provide consumer protection. Subsequently, they often exercise greater supervision of their employees so that compliance with weights and measures requirements is greater.

Another important activity that builds support for the weights and measures program is a public relations and outreach effort. Consumers rarely see weights and measures officials conduct inspections, so they may not know that the program exists. Consequently, weights and measures programs must make an effort to "tell their story" to make consumers aware of how weights and measures programs benefit the public. These same public relations efforts can educate consumers on how they can better protect themselves in transactions. A knowledgeable public also means that consumers with complaints will know who to contact to report a perceived problem in the marketplace. Many successful weights and measures investigations have been triggered by consumer complaints.

14.5   Strategic Planning

Another major administrative responsibility of the directors of weights and measures programs is to have a written strategic plan. The strategic plan should describe the program objectives, outline the strategies to achieve specific outcomes, and set intermediate milestones for the program to achieve the objectives. It should have programmatic measures that demonstrate progress toward each of the objectives. A concise statement of objectives, strategies, desired outcomes and milestones can be an effective way to educate policymakers and legislatures about the weights and measures programs and the problems that exist to achieve the objectives.

The strategic plan should be an honest assessment of the health of the weights and measures aspects of the commercial measurement system in the state. It should describe what is needed to maintain and improve the situation, the problems and obstacles that exist to achieve success, the strategies and resources needed to make progress, and the intermediate milestones that can be used to measure progress toward the objectives. A good strategic plan will be a valuable tool for managers when preparing budget proposals and allocating resources.

14.6   Management Responsibilities

The following are some of the management responsibilities of the weights and measures director:

- Weights and measures laws, regulations, test procedures and interpretations of weights and measures requirements are consistent with national standards and recommendations.

- Regulatory control is exercised in an efficient and effective manner. Alternative approaches have been considered, evaluated and decisions documented for the regulatory approach(es) used.

- Effective, fact-based budget and legislative proposals are developed and submitted.

- Metrology laboratory measurements are traceable, uncertainties are based on sound statistical methods, valid test procedures are used and the laboratory utilizes an effective quality system.

- The regulatory program maintains a balance of interests of industry, consumers and regulatory officials.

- Inspectors conduct inspections and tests using valid procedures, assist business owners to understand their responsibilities to provide accurate transactions and create a "culture of honesty" for business activities in the commercial marketplace.

- Inspection results are summarized, analyzed, benchmarked and used to assess program effectiveness and resource allocation.

- Inspector assignments are based on an assessment of the distribution of the population, the number and size of regulated businesses in an area, travel times between inspection sites, special knowledge and test equipment needed, length of time required to conduct effective inspections, and the scope of the inspector's responsibilities.

- Inspection assignments and scheduling are efficient and inspector performance is analyzed.

- Effective oversight of service companies exists to support the adjustment and repair in a timely manner of weighing and measuring devices to maintain them in an accurate and correct condition.

- The weights and measures program has an effective communication and interaction with businesses that are regulated and with the manufacturers of weighing and measuring devices so that stakeholder input can be obtained for proposed new regulations, changes in programs and priorities, and to obtain industry input on the costs and benefits of the proposed changes.

- Continuous communication and consultation with industry stakeholders and trade associations occurs to seek industry input when new programs are implemented, when changes are made to existing programs and to keep the regulated businesses informed of regulatory activities and changes in policy.

15.0   Program Funding and Fees

15.1   Issues in Obtaining Funding

Weights and measures programs rarely have enough funds available to operate programs at an optimum level. Elected officials often are averse to raising taxes to fund regulatory programs, so they may seek alternative ways to fund these programs through the imposition of fees. Weights and measures directors must review and revise program priorities and explore different approaches to weights and measures enforcement to be as effective as possible based upon the funds available.

Budget proposals should convey the importance of the weights and measures program and reflect the changing priorities of state and local government. Budget justification is an ongoing task since budgets usually must be approved every one or two years. The priorities of elected officials regularly change due to economic and political issues as well as changes in administrations at the state and federal level. Weights and measures programs are sometimes overlooked, because the activities focus on maintaining the infrastructure for the commercial measurement system and they are not highly visible. Nevertheless, weights and measures directors must educate policymakers and legislators on the importance of weights and measures programs and continually promote the successes and benefits of the program.

Funding may be allocated totally or in part from the general fund of a state budget. Some states have fared well in justifying increased expenditures on weights and measures programs, while others have suffered reductions and the threat of elimination. To reduce the threat posed from general fund budget cuts, many states have moved to a partial or total fee funding system.

The fee systems used in weights and measures programs have taken several forms:

1. A portion of the state tax per gallon (or liter) of gasoline and diesel fuel sold is dedicated for weights and measures activities;

2. The annual registration of measuring instruments (also called device registration);

3. The annual licensing of businesses; and

4. Inspection fees.

Each of these options will be discussed in sections below.

Potential problems associated with any fee system include the following:

- Since the inspections are funded by the fees, the selection of weights and measures activities may end up being driven by those activities that provide the greatest revenue rather than by programmatic needs;

- Businesses expect "services" from the regulatory agency for the fees paid;

- The fees may be set too low to cover the costs associated with the inspections;

- The fees may not be raised with sufficient frequency or by an amount sufficient to cover the costs of inspection as they rise from year to year; and

- Programs tend to leave large capital equipment purchases (such as truck scale test units, laboratory equipment, etc.) out of their annual budgets. Because weights and measures programs have high capital costs, line items for these types of equipment must be included in annual budgets with depreciation over the life expectancy of the equipment.

Furthermore, many weights and measures regulatory activities are not based upon the inspection and test of fee-generating areas. These regulatory activities may be cut back because weights and measures programs are obligated to conduct inspections of measuring instruments. States should therefore consider all of these potential problems when designing a fee structure.

The importance of strong legislative and administrative support for a strong weights and measures program cannot be overstated as factors that contribute to the success of the program.

15.2 Funding from State Engine Fuel Tax

Several states report that the apportionment of tax on retail sales of engine fuel has been fairly successful to funding a portion of weights and measures programs. One advantage of this approach is that little overhead or administrative cost is incurred by weights and measures programs to track these funds, since the tax departments of state government are already tracking the fuel sales data. A second advantage is the revenue source for the program is somewhat stable.

One limitation of this approach is that often the use of the funds generated by the sale of retail engine fuel is limited to testing liquid petroleum meters, which usually include retail fuel dispensers, loading rack meters and vehicle-tank meters. The remainder of the weights and measures program must still be funded by general revenue or another type of fee structure. Over the years, the quantities of gasoline and diesel fuel sold have increased, so the funds available for the petroleum inspection programs funded by this source of revenue have also increased. If the use of fuel becomes stable or decreases over time, however, the agency would need to increase the apportionment to prevent a shortage of funds.

15.3   Annual Device Registration

The annual device registration approach to fees appears to be the next most successful fee program in use today. Under this fee structure, the weights and measures program specifies which measuring instruments should be subject to registration fees and then receives authorization from the state legislature to set and impose the fees. Significant overhead costs are associated with this fee scheme, because the weights and measures program must maintain a record of each individual measuring instrument by company and location, add new measuring instruments as they are installed, delete those removed from service, annually mail invoices to the businesses for the fees, have a collection program for delinquent payments and take actions to collect the fees as necessary. The weights and measures official may also be assigned the additional responsibility of bill collector, which detracts from the primary responsibilities of the inspector.

The businesses that pay the annual device registration fees may also expect testing services, rather than regulatory inspections, from the weights and measures officials. This expectation usually allows weights and measures programs to conduct annual inspections of all registered devices in the state. If the weights and measures program must commit resources to inspection, it may not have the flexibility to assign resources to areas where greater oversight is needed.

Several weights and measures programs have added device registration fees to retail checkout scanners to generate funds to support price verification programs. This is another significant source of revenue if the fees are set at a level sufficient to support the inspection activities.

One state has reported successful use of device registration fees to support the weights and measures program for the inspection of measuring instruments. The program has been able to provide excellent supporting data to demonstrate the range and benefits of their activities. As a result, the legislature has established ranges for the registration fee for each type of measuring instrument, so the weights and measures authority may increase or decrease fees within the specified range to reflect the actual costs for inspecting the instruments that are registered, without requiring legislative approval for fee changes within the authorized range. The range of fees has been set at a level that is sufficient to cover the costs of the associated inspections. The industry supported the fee structure because they are more concerned about having fair competition among related businesses than they are concerned about the size of the registration fees. In this case, businesses believe that the fees are acceptable for the benefits gained from a more robust weights and measures regulatory program.

Additionally, general revenue funds may be provided for weights and measures inspections that do not involve measuring instruments, such as packaging and labeling inspections.

15.4   Annual Licensing of Businesses

Licensing of businesses is another potential source of revenue that can be used to cover costs of inspections. In addition to producing revenue, licensing provides control over businesses that do not meet their obligations, because the license can be revoked. A number of states use licensing fees to supplement their budget.

However, licensing presents several problems.

- Establishing equitable fees may be difficult.

- Licensing gives the administrative official quasi judicial functions in the matter of suspending and revoking licenses; if a license law does not give that power to the official, justification for the system fails.

- The administration of the system entails considerable work on the part of those in authority and introduces some complications for the licensees.

15.5 Inspection Fees

Charging fees for inspections is another option for a regulatory program. Many states charge inspection fees, and have found this approach to be a successful method for maintaining revenues. The revenue derived from inspections increases along with the amount of work carried out, unlike some other methods of funding.

A common pitfall is that inspections may be conducted for the purpose of collecting fees, rather than being conducted to determine compliance with weights and measures requirements. Additionally, weights and measures officials may rush through inspections, instead of conducting thorough inspections, because of the need to generate revenue. An inspector must balance inspections with other important activities, such as investigating compliance problems. It is important that weights and measures directors ensure the focus of inspections does not shift from enforcing the law to generating revenue.

16.0 Fines and Penalties

Weights and measures programs typically strive to achieve business compliance with weights and measures requirements using the lowest level of regulatory action possible. However, sometimes businesses do not perform at acceptable levels despite repeated warnings and low levels of enforcement actions. Therefore, weights and measures programs must have the authority to impose or pursue higher levels of enforcement action through the use of fines and penalties.

Historically, weights and measures programs have had the authority to pursue criminal prosecution through the county district attorneys, which could result in fines and even incarceration of business owners. However, many district attorneys have heavy caseloads dealing with other serious criminal offenses, so some do not place high priority on weights and measures criminal cases. Furthermore, criminal cases for weights and measures violations are relatively rare events, so the district attorneys frequently are not familiar with the weights and measures laws under which the case must be prosecuted. As a result, few weights and measures violations may go to criminal court.

In recent years, many weights and measures programs have been authorized to issue civil penalties on businesses for weights and measures violations. Weights and measures programs that have the authority to issue civil penalties have found them to be very effective and they have a much smaller administrative burden than criminal prosecutions. However, businesses must be offered a reasonable process to challenge civil penalties imposed by the regulatory agency in the event they find the penalty inappropriate.

Because of the rather immediate impact of civil penalties, it is imperative that weights and measures inspectors conduct proper inspections following appropriate procedures and using appropriate physical standards and technical regulations. It is critical that all aspects of the inspection process be valid to maintain the credibility of the weights and measures regulatory program.

When fines and civil penalties are used as a tool to gain compliance, the fines must be of sufficient magnitude so that they serve as an effective deterrent to noncompliance in measuring instruments or marketing practices. Because weights and measures regulatory activities are fundamental to the economy, weights and measures laws have existed for a great many years. Consequently, many of the laws that authorized the amounts of fines or penalties for violations of weights and measures requirements are outdated and do not serve as effective deterrents to noncompliant activities. Weights and measures fines and civil penalties must be reviewed and updated periodically to maintain effectiveness.

Fines and penalties should be used exclusively as tools to gain compliance with weights and measures requirements. The following should be examined to determine if fines and penalties are being used inappropriately to generate revenue, undermining the integrity of the regulatory program.

- The percentage of violations for which fines are imposed should be tracked and benchmarked against those of other states. An unusually high percentage of violations that result in fines may indicate that fines are being used for revenue generation.

- The amount of money generated by fines for each inspection discipline should be compared to the total budget of the weights and measures program allocated to the inspection discipline. A high percentage may indicate that fines are being used for revenue generation.

- If the amount of revenue generated by the weights and measures program is used as the basis for budget justification or to justify adding positions to the program, the program may have lost its focus on using fines as a tool for compliance. Many states require any fines that are derived from non-compliance be dedicated to other activities in the state budget, which reduces the incentive to use fines for revenue rather than compliance.

Weights and measures directors often find themselves tempted to use generated funds as a budget justification because it may be to explain and substantiate their program to legislators and auditing officials who may not understand the complexities and benefits of the weights and measures program. Furthermore, state legislatures usually are more concerned with prioritizing and balancing the state budget. Thus, directors should focus on educating the legislators and justifying the expenditures.

17.0   Inspector Positions and Responsibilities

Each state must decide on the most efficient way of assigning responsibilities to weights and measures officials. Because towns and cities are distributed throughout the state, inspectors are usually assigned a particular part of the state as their area for inspection. The inspector assigned to an area performs inspections in most of the different inspection disciplines within the assigned area and typically lives within the area to which he or she is assigned so that travel time and costs are limited.

Ideally, the weights and measures official should be a full time official, because the inspector must be knowledgeable in many different laws and regulations, measurement areas, measurement technology, understanding and utilizing electronic audit trails, retail and marketing practices, and distribution systems that support the commercial measurement system.

The list of weights and measures disciplines shown previously in Table 2 provides an indication of the breadth and depth of knowledge and inspection skills needed by weights and measures officials. The use of complex software in measurement applications further complicates the inspection of measuring instruments. Additionally, many jurisdictions use laptop or tablet computers to capture and report inspection results, so the inspector must be familiar with the software and may be required to update database records when new or different businesses and measuring instruments are found.

The inspector is also required to have a wide range of technical knowledge in order to understand the transactions that are being regulated. New inspectors should be required to master the test procedures and enforcement policies of the state before the inspector is permitted to conduct unsupervised inspections. The inspector must understand the certification and traceability aspects of the standards used to test packages and measuring instruments.

As the inspector gains experience, he or she should learn the interrelated aspects of the commercial measurement system in each inspection discipline to appreciate the ramifications of regulatory actions. The inspector should have a basic knowledge of physics to understand the operation of the many measuring instruments that are tested and inspected, to understand how the test procedures determine the operating characteristics of the measuring instruments under test, to understand corrections that must be made during the test process, and to understand the effects that environmental factors can have on test results.

The inspector should have a basic knowledge of statistics to understand how test results may vary due to random and systematic effects during the test and measurement processes. Analytical skills are required to understand the meaning of the test results and to put the test results into the proper perspective, so that appropriate regulatory action can be taken.

The inspector must also have good self discipline and management skills, because most inspectors work with a high degree of independence, and must plan his or her own work schedule and work assignments in an efficient manner. Finally, it is important that inspectors have good people skills and common sense so that inspectors can work effectively with business people and their employees to achieve compliance with weights and measures requirements while striving for cooperation and corrective action as necessary. The inspector must be empowered to use this knowledge and the regulatory authority to take the enforcement action necessary in each situation to best achieve compliance with the weights and measures requirements.

Some weights and measures inspection disciplines are typically assigned to a weights and measures specialist. A specialist is an inspector who is assigned to perform one or two particular types of inspections because special equipment is required, special knowledge or skills may be needed, and there may be safety concerns that are best addressed by someone specially trained in these safety techniques.

For example, large vehicle scales test units are typically large trucks that carry 20 000 lb or more of test weights, which may include a weight cart that must be loaded with test weights and driven across the scale. The weights and measures inspector may be required to have a commercial driver's license to drive the vehicle. The truck carrying the test weights also has a crane or hoist for unloading and loading the weights, for which safety training may be required.

Another common specialist position is the inspector of the test unit for truck-mounted LP gas meters. A special prover is needed because LP gas does not remain in liquid form at normal atmospheric pressure. Consequently, the LP gas prover is a closed prover capable of withstanding pressures up to at least 200 psig. LP gas has a large coefficient of cubical expansion, so LP gas is usually required to be sold on a temperature compensated basis. Because of the variables involved in the test, the inspector must make temperature and pressure corrections to the prover capacity, make temperature corrections to the LP gas liquid, and use appropriate safety techniques to handle a flammable liquid that can also cause skin injuries (from freezing) if the LP gas liquid or vapor spills on to or sprays on to any part of the inspector's body. The prover itself is large, so towing a trailer with an LP gas prover behind a car or truck may require additional training, skill and experience.

In some areas, weights and measures officials are assigned additional responsibilities beyond those of weights and measures, such as food inspection, egg inspection, fertilizer sampling, fuel sampling, sign posting inspections, or any number of other inspections that may be conducted in businesses that the weights and measures official must routinely enter to conduct inspections. The addition of inspection and regulatory responsibilities beyond those of weights and measures should be avoided because the inspector is already facing considerable demands in having to be knowledgeable on weights and measures laws, regulations, and measurement technology, and then be expected to be an expert in one or many more regulatory areas.

The following points should be considered in staffing a weights and measures program:

- A well-trained and properly supervised group of personnel is essential for success in weights and measures programs.

- When possible, full time employees should be assigned to weights and measures duties. The work is highly technical, and study and experience on the part of the official are required for efficiency. Combining weights and measures activities with business pursuits can pose particular problems, including conflicts of interest.

- In a city of significant size, weights and measures work can occupy the time of several full time officials. In other less populated areas, a county subdivision or several counties may be required.

- If state law requires the appointment by local authorities of separate weights and measures officials in *all* political subdivisions of a certain class—cities, counties, etc.—without regard to the amount of work in those subdivisions; the local jurisdiction that is required to appoint should be large enough to support a full-time official. In some states that have not provided for state inspectors to take care of thinly settled sections, an attempt has been made to meet this situation through legislation permitting adjoining jurisdictions to combine for purposes of weights and measures supervision to appoint an official to serve jointly for such jurisdictions. However, in many such cases, if the jurisdictions are not required to combine, they do not do so and, therefore, the problem remains.

- In the case of activities requiring specialized equipment such as the testing of vehicle scales or the calibration of vehicle tanks, an exception should be made to the recommendation that all work be performed by one individual. In these fields it will ordinarily be neither economical nor efficient to provide the relatively expensive special equipment for each inspector, and the work is best carried on by separate crews trained for and concentrating upon their particular specialties.

The assignment of inspection areas to officials must consider a number of factors. Some of these factors are listed below.

- Population and distribution of towns within the state;

- Number of businesses and measuring instruments that must be inspected in each inspection area;

- The topography and time needed to travel between businesses and towns to be able to conduct inspections;

- The time needed to conduct each type of inspection for the disciplines being inspected;

- The equipment needed to conduct each type of inspection;

- The characteristics of each type of measuring instrument as they relate to maintaining instrument accuracy between times of inspection;

- The availability and use of private service companies to test and maintain measuring instruments in each inspection area;

- The number of transactions; and

- The experience and abilities of the inspector.

The rationale for the decision to assign inspectors specific inspection areas, sizes of areas and the creation of specialized inspectors should be documented so that as conditions change, the basis for decisions may be reviewed and revised as necessary. This information is needed to justify position descriptions for general and specialized positions. The documentation for the assignment of inspection areas may also serve as a basis for reviewing the need for additional resources or the reallocation of existing resources as populations and business concentrations change.

17.1   Ensuring Accuracy in Instrumentation Testing

One of the responsibilities of weights and measures officials is to test measuring instruments. This section explains the factors that can affect the accuracy of such testing. The degree to which the test of a measuring instrument duplicates the conditions of actual use varies from instrument to instrument. For example, the tests of retail fuel dispensers and loading rack meters closely simulate the normal conditions of use, because wind and rain do not affect the measurement. However, weights and measures officials usually do not test fuel meters in very cold or very hot temperatures, although businesses use these metering devices under all of these conditions. Sometimes scales are used under conditions that typically are not duplicated in the test process. For example, a vehicle scale may weigh all vehicles from pickup trucks to semi-tractor trailers. The distribution of the load generated by a truck on a scale may be significantly different from truck to truck based on axle configuration and from the load distribution when the scale is tested using field test weights. If the weights are placed directly on the scale platform, then the mass of the weights is distributed over a relatively large area of the platform and, in a complete test, placed over each section of the scale in succession. While the loading pattern does not exactly duplicate the conditions of use, this may be the most reasonable way to test the scale. If the vehicle scale test unit has a weight cart, then the distance between the axles and between wheels on the weight cart determine the loading points of the scale platform.

Weather conditions also may affect testing of measurement instruments. For example, it is difficult to test a vehicle scale located outdoors on a windy day. The effects of the wind on the weights and on the platform usually cause the weight display to vary to such an extent that accurate reading of the scale indication cannot be done. However, vehicle scales continue to be used on windy days, during rain and snow, and even when ice may cause the scale platform to bind.

The scale operator should take care to obtain the most accurate weight reading possible under the weather conditions. If the scale is covered with ice, the scale operator should break the ice away from the edges of the platform to eliminate binding before the scale is used for a commercial transaction. It should be apparent from the conditions under which some scales are used, that the accuracy of actual commercial transactions may not be within the tolerance for the scale during the test of the scale under more ideal conditions. Hence, weights and measures officials should be concerned about the accuracy of transactions and not limit their regulatory efforts only to the testing of the measuring instruments.

Handbook 44 states that measuring instruments must be accurate under normal conditions of use. Each weights and measures program determines how they control the use of measuring instruments under conditions that are likely to result in measurement errors during transactions.

In general, weights and measures officials should adhere to the following when testing measuring instruments.

- Scales and meters may be tested under any environmental conditions in which the devices are used for transactions, yet still produce accurate test results.

- Scales and meters shall not be tested when weather conditions interfere with the measurement process to the extent that valid test results cannot be obtained.

- Standards shall be stored and transported in appropriate environmental conditions so that the accuracy of the standards will be maintained when not in use.

18.0   State and Local Weights and Measures Jurisdictions

Some states have local (city and/or county) as well as state weights and measures officials. The local jurisdictions have the authority and responsibility to enforce weights and measures laws in the same manner that state officials have to enforce state laws. While local jurisdictions may have the authority to issue their own regulations or ordinances for weights and measures, the local weights and measures requirements, inspection procedures, and enforcement actions should be consistent with the state requirements. The importance of consistency at the state and local level is the same as the justification for weights and measures requirements and procedures to be consistent from state to state.

The use of state-only versus state and local jurisdictions for weights and measures enforcement each has advantages and disadvantages. Discussions at workshops for state and local weights and measures administrators in 2004 provided valuable perspectives on this subject. Generally, local weights and measures programs tend to have a higher ratio of inspectors to the number of businesses and commercial measurement instruments than state programs. Consequently, local programs often provide a higher level of service than most state programs. This arrangement is often justified because the large number of businesses in cities and towns has great economic impact on the surrounding region. City jurisdictions usually have shorter travel time, so they can spend more time on inspections. Heavy traffic and the lack of parking near local businesses may increase travel time, but that situation will exist for both state and local inspectors when they must conduct inspections in urban areas.

Local officials usually respond faster to complaints, and do their own re-inspections of stores and measuring instruments after the instruments have been rejected by the inspector. The shorter travel distances make this feasible. State offices often rely on service companies to put measuring instruments back into service, because they usually cannot afford to incur significant travel time to conduct an official test to put instruments back into service after a rejection.

One concern expressed in the NIST administrator workshops was that some local jurisdictions do not coordinate their activities with the state weights and measures office or from jurisdiction to jurisdiction. Another concern was that some local officials may not follow the instructions, guidance and test procedures specified by the state offices, leading to inconsistencies in procedures, practices, and enforcement policies. Some cities and towns may not have enough work to justify a full-time weights and measures inspector, so the inspector may be assigned other responsibilities or the weights and measures position may be a part-time position. Local and part-time inspectors may not get the training they need on a regular basis to keep up with technology and changes in the marketplace. Additionally, the part-time inspector may not be able to maintain proficiency in the many test procedures for the wide range of inspections that should be performed.

Some local jurisdictions have overcome these problems by providing services to two or more cities or towns. This solution allows the local weights and measures program to justify full-time positions and maintain the proficiency of inspectors in a variety of inspection disciplines. Furthermore, expensive test equipment often can be more effectively utilized when the inspection responsibilities for several local jurisdictions are under the authority of a single local weights and measures office. Sometimes two or more local jurisdictions share the more expensive test equipment to reduce the cost to each individual jurisdiction

Another approach to more effectively utilize expensive test equipment is for local jurisdictions to perform only those inspections for which they have a sufficient workload. They may establish an agreement with the state weights and measures office to test measuring instruments that require specialized equipment and which take more time to test, such as vehicle scales and truck-mounted LP gas meters.

States may have a requirement that cities above a certain population level must have their own weights and measures program or the state will charge the city for providing inspection services. Over the years, the number of local weights and measures programs has decreased, as cities have opted to have the state conduct weights and measures inspections within the city. Often city officials view the dropping of the local weights and measures program as a cost-cutting measure. Unfortunately, many times when a state program assumes the inspection responsibilities of cities, the state program does not get any increase in funding or personnel to do the extra work. The fees charged to the city may not cover the cost of the state inspections. The effect of these actions is that the number of inspections conducted in the cities may be reduced.

19.0   Complaint Investigations

The investigation of complaints from consumers or from businesses about alleged measurement problems or unfair marketing practices is part of the normal operation of a weights and measures regulatory program. The most common consumer complaint is a problem they believe exists regarding a purchase that was made. Complaints must be taken seriously by weights and measures programs and they should receive a high priority for investigation.

Weights and measures officials should encourage the public to report complaints, and should investigate them carefully. An effective educational program conducted through a public relations program will help reduce unjustified complaints and improve the quality of information received for those that are well founded. Although some complaints are unwarranted, weights and measures officials should investigate each one, because some complaints that initially appear questionable may result in the discovery of serious violations.

When a complaint has been found to be justified, a variety of actions are possible. If the complainant has suffered damages, the official may be able to arrange a settlement. In considering whether prosecution is called for, the official should consider the offender's previous record, the severity of the offence, and other factors. Each case needs to be evaluated individually.

The weights and measures official should always take steps to prevent a recurrence of the problem. The appropriate action depends on whether the event was localized or pervasive. If the problem runs throughout an industry, it is usually necessary to enlist the cooperation of other jurisdictions, either to collect further information or to apply corrective measures.

After a complaint has been investigated, a formal report should be provided to the complainant, regardless of the outcome. In general, the results of investigations may not be suitable for publicity. Care should also be taken not to release any information prematurely, because doing so might jeopardize a successful outcome by alerting the subjects of the investigation to the intentions of the official.

While many consumer complaints cannot be substantiated as a result of investigation, a significant number of complaints can be verified, and often uncover problems or fraudulent practices that are not found in routine inspections. For example, California county and state weights and measures officials received several complaints about a few service stations for short-measure deliveries. An extensive undercover investigation revealed that a few unscrupulous people were modifying the software for gasoline dispensers so that they would deliver short-measure for quantities other than at five and ten gallons, which were the points at which the inspectors normally test the accuracy of the dispensers. In another instance, gasoline dispensers were programmed to deliver short-measure, but when the station attendant saw the weights and measures inspector arrive to conduct an inspection, the attendant would turn the power off to the dispensers to reset the software and deactivate the fraudulent software until after the inspector left the station.

Measurement errors may be the result of carelessness or a lack of training of the operator of the measuring instrument. Because weights and measures officials inspect and test measurement instruments on a rather infrequent basis, complaints from consumers can be valuable sources of information to alert inspectors to measurement inaccuracies.

20.0   Scheduling Work Assignments

Many weights and measures inspectors are assigned a particular area of a state and perform inspections of all businesses within that area. Based upon the priorities established by the weights and measures office, the inspector knows the types of businesses, measuring instruments, and transactions that are to be checked at a particular time of the year. Some inspections must be performed outdoors, such as testing retail fuel dispensers and vehicle scales, but others, such as supermarket inspections, are done indoors. The inspectors understand that if gas stations and supermarkets are to be inspected on the same day, it is reasonable and considerate to perform supermarket inspections before the service station inspections so that the odor of gasoline from the inspector's clothes does not offend shoppers in supermarkets or raise concerns regarding sanitation if the inspector must handle food products.

The procedures for scheduling inspections vary considerably among different offices. In some cases, inspectors are given limited direction on when or which inspections to conduct at a given time. The inspector may simply be aware that it is his or her responsibility to inspect all regulated businesses and test all commercial measuring instruments within the assigned area on an annual basis or within the time period prescribed by the weights and measures office. A state weights and measures program may require inspectors to submit work itineraries at least one week in advance. In other cases, the weights and measures office may assign businesses for inspection during a particular time period. Either way, the inspector still has the flexibility to change the itinerary as necessary based upon work issues, investigation of complaints, and weather conditions. However, weights and measures directors have often reported that the use of work itineraries has increased the efficiency and productivity of inspectors.

When work itineraries are prepared, the weights and measures office may send preprinted forms to the inspector to reduce the time the inspector must spend filling in routine information. Ideally, this information is entered electronically and is available to the inspector on a laptop or handheld device. Frequently, the inspector will record the time of arrival and departure at a business so the time required to conduct the inspection is documented. Additionally, the inspector usually records travel time on the weekly work report so that the office staff know how much time the inspector expended on each type of activity. The weekly work report also reflects any leave, vacation, sick leave, and holidays that the inspector has taken. This information is helpful in analyzing workloads and monitoring progress on assigned tasks. It also serves as a basis for comparing the performance of one inspector to another.

21.0   Knowledge and Training

One goal of the weights and measures program is to achieve uniformity among the different inspectors in all of their inspection activities. New inspectors must learn a wide variety of inspection procedures in a large number of disciplines. Inspectors are expected to become experts in their profession by learning from their experience and from the expertise of others with whom they come into contact.

Officials should not deviate from established procedures by taking short cuts in the inspection of businesses and the testing of measuring instruments. Test procedures should be updated as technology changes so that field tests are efficient, yet adequate, to verify performance characteristics. Changes to the test procedures should be documented and validated so that only appropriate changes are made, approved, and distributed to all parties (e.g., field inspectors and service technicians) that need the updated procedures.

For example, a weights and measures program may follow the inspection and test procedures contained in the examination procedure outlines of NIST Handbook 112 or the state may have its own inspection and test procedures. If a state develops its own procedures, the procedures must still be valid, adequate and sufficient.

The weights and measures program should have a defined training program that describes how new inspectors are trained in each discipline. Furthermore, a defined training process must be developed for updating new and experienced inspectors when the test procedures or the technical requirements change, which happens on an annual basis with the update of NIST Handbook 44. Documentation should be on record which indicates that these training sessions have taken place and which identifies the recipients of the training.

Written inspection and test procedures are essential to provide a common basis for inspection and test. Each inspector should have the current laws and regulations, inspection and test procedures, enforcement policies, handbooks and other reference material. A mechanism must be in place to permit the inspector to determine if measuring instruments found in the field have NTEP Certificates of Conformance.

A separate training procedure is not needed for each type of training. However, a comprehensive written procedure should be prepared that describes how training is to be done and how inspectors are updated on new measurement technology and marketing practices, and the inspection procedures and policies that go with them. The management must be able to demonstrate that these procedures are in place and that they are used.

Although written enforcement policies exist, the inspector must be able to assess each situation encountered in the field to determine whether or not additional investigation is needed and whether or not extenuating circumstances should affect the enforcement action to be taken. Sometimes the inspector may and should deviate from written policy, but these actions should be approved by his or her supervisor. The field inspector must think before taking action. Ill-advised action taken as a result of a lack of understanding of the situation or incomplete testing can have serious ramifications on the regulatory program.

Enforcement action taken by each inspector should be consistent for the same types of violations under similar circumstances. Documented enforcement guidelines should exist and be followed for normal types of violations.

Electronic audit trails are relatively new in measuring instruments, but they are often underutilized as a tool by the weights and measures officials. Because access to audit trail information is not standardized, some weights and measures officials do not check the audit trail information. Furthermore, the inspection reports may not have an area to record the "counts" contained in audit trails, so the inspector does not have the information available for subsequent inspections. If a measuring instrument has an event logger, the inspector may not know how to print out the information and may not take the time to study the audit trail entries to determine whether or not the measuring instrument calibration or features are being used fraudulently between inspections

Providing information on audit trails to the field inspector should be a priority of the weights and measures office. Weights and measures offices are encouraged to provide their inspectors with a file of the information from NTEP Certificates of Conformance on how to access audit trails.

Supervisors should work with each inspector on a regular basis. The supervisor should observe how the inspector interacts with business owners and managers to see whether or not any problems appear to exist. Additionally, the supervisor can observe the inspection and test procedures used by the inspector and correct any deviations from prescribed procedures. The supervisor benefits from this process by experiencing firsthand the difficulties of conducting field inspections and seeing new measuring instruments, technology, and methods of sale in the field. This knowledge can be used to update policies, procedures, standards, and enforcement practices not only within the state, but on the national level. Again, the effort to strive for national uniformity benefits regulators and businesses alike.

A major challenge in the area of weights and measures is the ability to provide the necessary training to inspectors to develop their knowledge and professional skills. Because of the complexity of weights and measures, no one is an expert in all of the many inspection disciplines. It is difficult to find experts in different inspection disciplines to provide training to the many weights and measures officials across the country. In addition, each class may have learners at different levels of expertise, so efforts must be made to keep the course interesting and informative to more experienced officials, while providing the necessary guidance to those with less knowledge.

Another training concern has been the lack of up-to-date training material available to state and local weights and measures programs to use in their own training programs. However, updating training material is a high priority for the NIST Weights and Measures Division (WMD) and some training material is available on the WMD web site.

The training needed for weights and measures officials can be placed in two categories; training for new inspectors, and advanced training. A new inspector needs training in the basic laws, regulations, measuring instruments, measurement technology, measurement applications, and inspection techniques to enable to the inspector to conduct independent inspections consistent with accepted test procedures, policies and interpretations of weights and measures requirements.

A training program for inspectors is encouraged to have the following components:

- An adequate, uniform, and defined training process that is completed within a specified time period.

- Minimum competencies are defined and include an understanding of the following:

    o The largest scale or meter division appropriate for each area of inspection;

    o The "normal" or average quantity of a transaction through a measuring instrument (relating to selecting the appropriate measuring range of an instrument);

- o The maximum quantity of a transaction that can typically be expected in each area of inspection (relating to the appropriate capacity of a device);

- o The minimum quantity for transactions that can typically be expected in each area of inspection (relating to the minimum quantity for which a measuring instrument is designed to weigh or measure); and

- o The typical types of measuring instruments that are routinely used for transactions in each area of inspection.

- Adequate supervision and oversight of field inspectors.

- Knowledge requirements (including the extent of detailed knowledge required for new and experienced inspectors), skills and abilities are defined for each inspector position.

- Training on defined test procedures, enforcement policies and agency interpretations in their field inspections and enforcement actions.

- Written tests, proficiency testing and observation of performance of the inspector in the conduct of field inspections.

- Specialized advanced training for experienced inspectors.

- Periodic planned training and testing for experienced inspectors are required to maintain skills.

- Ideally, a certification program formalizes the competence of inspectors.

More experienced officials develop a comprehensive understanding of many aspects of weights and measures, including:

- The principles of weights and measures enforcement;

- The operation of the commercial measurement system;

- The ramifications of weights and measures enforcement on businesses and the distribution of products;

- The roles of the many stakeholders in the commercial measurement system, including state and federal agencies, regulators and regulatory jurisdictions;

- The measurement technologies used in measuring instruments;

- The design of measuring instruments and the concept of tolerance application over a range of influence factors, short-term and long-term repeatability of measurements, and the tradeoff of cost versus accuracy in terms of the measuring instruments and the labor required to make measurements;

- The physics of how measuring instruments work and the variables that affect the accuracy of measurements and transactions;

- The physical standards used to test measuring instruments and to check the net content of packages;

- The factors that affect the repeatability of test results during the enforcement testing of commercial measuring instruments;

- The statistical concepts that apply to package inspection, risk-based inspections and the analysis of test results for measuring instruments;

- The marketing practices used in different segments of the commercial measurement system; and

- The intent of laws, regulations, and standards and how to apply these concepts to the wide range of measurement and marketing practices found in and developed for the marketplace.

Weights and measures officials and representatives from industry who develop this extensive knowledge routinely become leaders of the weights and measures community. Directors and supervisors must have the technical knowledge and the understanding of the commercial measurement system to provide the technical direction to field inspectors, but to be effective leaders, they must also have the management and people skills.

It is important that the individuals who excel in both technical knowledge and management skills dedicate some of their time to participating in regional, national and international weights and measures (legal metrology) meetings. Through their participation, these individuals can help guide the development of new regulations and standards so they are technically sound and balance the interests of all stakeholders.

22.0   Evaluation of Inspector Performance

The objectives of weights and measures enforcement can be easily stated, as was done at the beginning of this document, but the evaluation of inspector performance is an effective way to convey the importance of these objectives to the inspectors and to provide feedback to each inspector on his or her contributions to meet these goals. Data needed to monitor and evaluate inspector performance should be collected and tabulated.

The evaluation must consider the types of devices within the inspection area for each inspector, the population density and distribution, and the types and number of businesses within each inspection area. Other factors may also be significant to the evaluation and may vary from state to state. As mentioned earlier, the weights and measures program should track at least the following information to evaluate inspector performance: percentages of time spent doing different types of inspections, travel time, annual leave (vacation), sick leave and holiday time.

Below are examples of some of the information that may be tabulated and graphed to evaluate inspector performance.

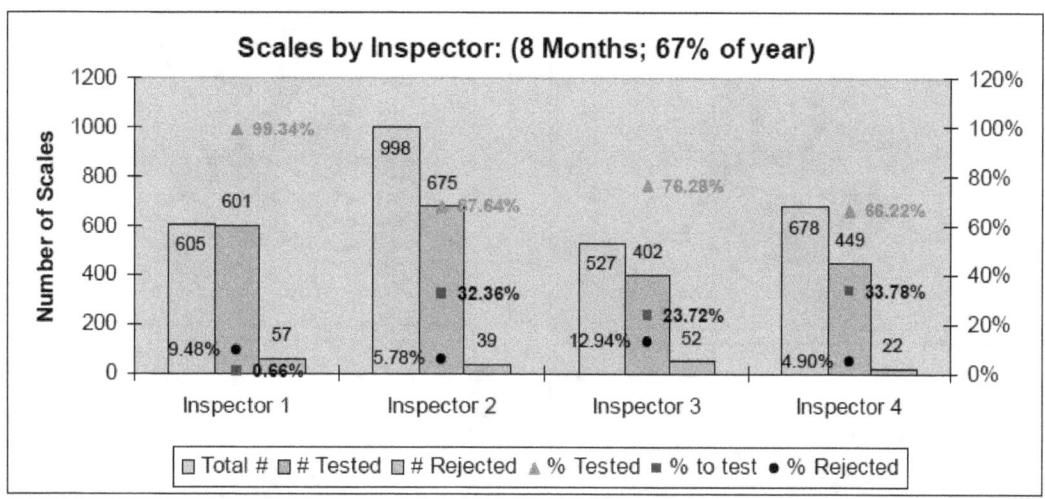

Figure 2. Inspector Performance Evaluation – Scales

Figure 2 shows performance by four different inspectors. The chart indicates the total number of scales, number and percent of scales tested, and the number and percent rejected.

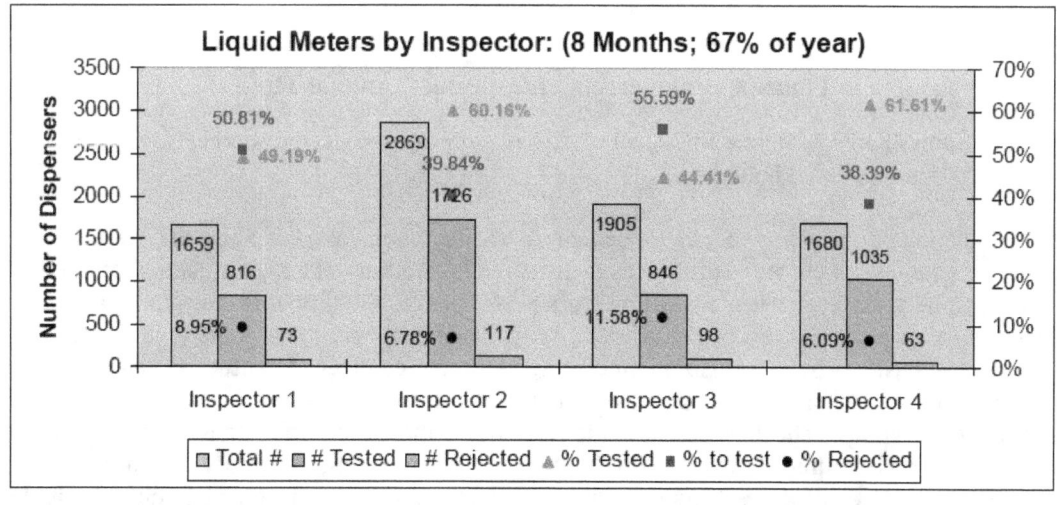

Figure 3. Inspector Performance Evaluation – Liquid Meters

Figure 3 shows similar data for liquid meters. Note that overall patterns for each of the inspectors are similar, although one inspector tested a much greater number of meters than the others.

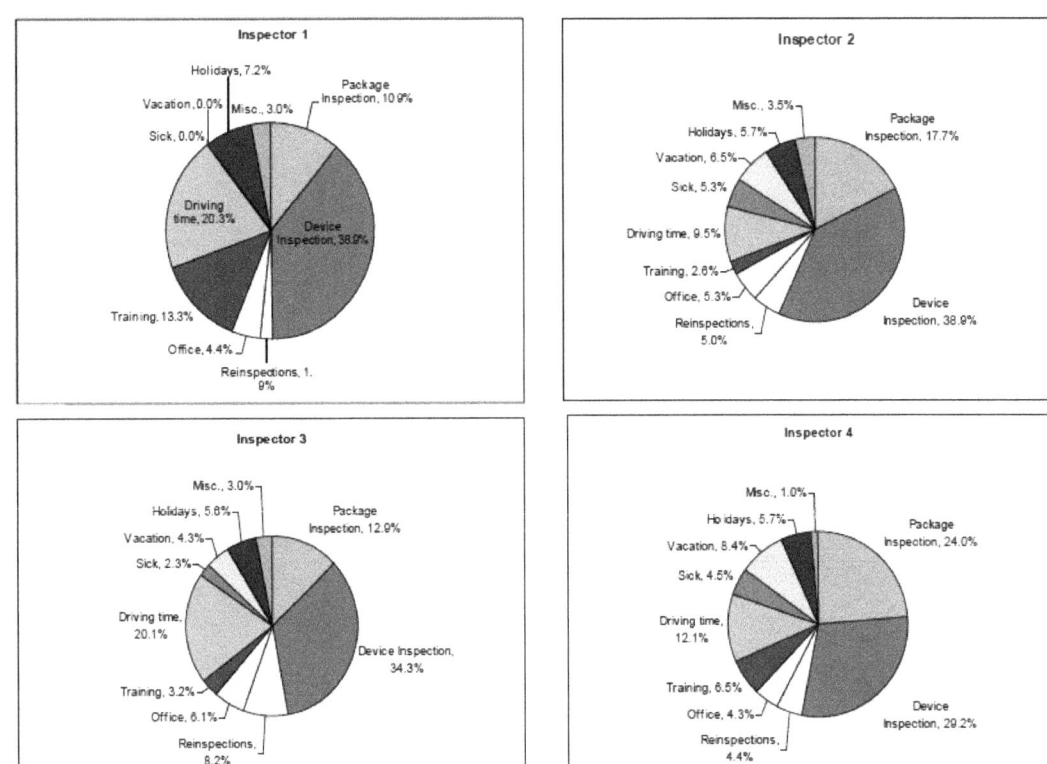

Figure 4. Comparison of Inspectors' Annual Time

Figure 4 shows how much time was spent of different types of inspections, and also time for vacations, sick leave, and holidays.

The evaluation of inspectors should not become a "number crunching" exercise that results in inspection quotas. There are many reasons why some inspections take longer than others, and these must be considered when evaluating work performance. Quality of inspection as well as quantities of inspections must be considered. Some inspectors may spend more time on the education of business owners regarding the laws, the proper conduct of transactions and the business owner's responsibility to maintain accurate measuring instruments and to conduct accurate transactions. The distribution of the types of businesses in different assigned inspection areas also affects the time allocated to different types of inspections. Obviously, the driving time between towns and businesses in rural versus more urban areas also affect the amount of time the inspector can spend performing inspections. Analyzing productivity and performance statistics is a complex process, but collecting and using the inspection data are required before this important management function can be performed.

23.0   Alternative Approaches to Regulatory Control

A major aspect of any weights and measures program is choosing an approach for inspection that will result in the highest level of compliance given the resources that are available. As part of this decision process, the director must determine the highest priority areas for inspection to maximize the benefit of expended resources. In the past, many legal metrology programs focused on testing all commercial measuring instruments every year. The theory was that if the devices were accurate, then the transactions would be accurate. Unfortunately, since a variety of errors (both accidental and intentional) may occur in the transactions, accurate devices are not sufficient to ensure accurate transactions. Consideration must be given as to the best way to ensure accurate transactions based upon the manner in which business is done. When considering alternative approaches to regulatory control, one must appreciate that the visibility of the weights and measures inspector performing regulatory inspections is an important influence in stimulating businesses to operate honestly.

Several approaches may be considered for weights and measures regulatory control:

- Testing 100 % of the measuring instruments on an annual (or other specified) basis;

- Variable frequency inspections;

- Risk-based inspections and statistical sampling that consider inspection data, past compliance rates for measuring instruments, company compliance history, and company service programs in effect;

- Utilizing inspection results from private service companies; and

- Delegating inspection responsibilities to private companies.

A weights and measures director may find some combination of these approaches works best for their program. A program may also use tools such as marketplace surveys and undercover test purchases to augment their approach.

23.1   100 % Device Inspection

The approach of testing 100 % of the measuring instruments in a given segment of the commercial measurement system is advised when a segment of the commercial measurement system has never (or not recently) been subject to weights and measures inspection. In this situation, one must be sure that the measuring instruments are accurate and correct before other problems can be addressed.

Weights and measures inspections in the United States have often been based on testing every commercial device and every retail business periodically to ensure that the owners are fulfilling their responsibility to maintain weighing and measuring devices so they comply with specifications and performance tolerances. In the past, the goal was to have an "official" test (that is, a test by a regulatory official) of every commercial weighing and measuring device on an annual basis. In some jurisdictions, this test cycle is specified in the state weights and measures law.

Over the years, many jurisdictions have not been able to perform device inspections on an annual basis. In many cases, the reality is that the test cycle has been extended in spite of any time requirement specified in the law, because sufficient resources are not available to inspect all commercial measuring instruments on an annual basis. In some cases, weights and measures directors have effectively used the mandated test frequency to obtain additional funds to increase staff or obtain equipment to increase the efficiency of inspections to complete the annual testing of all devices.

In other cases, when the weights and measures program brought the issue to the attention of the state legislature that they were unable to perform inspections on all devices within the time frame specified in the law, the legislature changed the law to extend the inspection period for official inspections. The high compliance rates for accuracy and specifications requirements for retail motor fuel dispensers (often greater than 95 % and 90 % respectively) often cause legislators to believe that cutbacks in the frequency of inspections can be made without significant consequences in compliance.

Unfortunately, the decrease in compliance often occurs at a slower rate than a cutback on inspections, so the consequences of reduced compliance due to decreased inspections may not be immediately apparent. However, this effect is more readily evident when jurisdictions have had to completely discontinue an inspection activity due to budget cuts, but were able to restore the program after a few years when funding became available. Experience has shown that the decrease in compliance is significant compared to the compliance level that existed prior to the discontinuation of the program.

The approach of 100 % device inspection is often the best approach for the inspection of measuring instruments when:

1. The rejection rate for a particular type of measuring instrument is high, because the accuracy is not likely to be maintained over a time period greater than the official test cycle. This may be due to the conditions of use, a harsh work environment for the device, the product being measured causes extensive wear on the device, owner/user fraudulent manipulation of the device, or other factors.

2. Weighing or measuring devices are used on a seasonal basis and the bulk of the measurements are made in short time.

3. The owner/user does not maintain the accuracy of the measuring instruments and does not comply with all technical regulations.

4. There is not an adequate scale or meter service industry to provide regular private test services to the device owners.

5. The calibration of a device may need to be changed on an annual basis due to the characteristics of the product being measured. The best example of this is grain moisture meters where the calibration of the meter is affected by the biological characteristics of the grain.

6. The product being measured is relatively expensive and the costs of measurement errors on transactions is significant or the volume of product measured is high and of great economic value.

7. The jurisdiction has sufficient personnel, adequate test equipment, and resources to effectively perform the inspections within the necessary time frame.

23.2 Variable Frequency Inspections

Variable frequency inspections should be used when a state has a database of inspection results over a considerable period of time. In this instance, the program would adjust its inspection efforts and frequency of inspection based upon the inspection history by type of business, retailer, type of measuring instrument or type of packaged products. The concept is that companies or measuring instruments that have had high compliance rates over an extended period of time do not need to be inspected as frequently as those that have lower compliance rates. With this approach, the weights and measures director reallocates resources into areas with low compliance rates or into areas that have not been routinely inspected. The result is that the weights and measures resources are used more effectively by targeting problem areas.

This approach works best under the following conditions:

1. The weights and measures agency has a database of all the businesses and measuring instruments in use within the jurisdiction.

2. The compliance levels for some types of businesses or measuring instruments are high.

3. Businesses with measuring instruments routinely obtain service from private service companies to test and maintain their instruments in compliance with weights and measures requirements.

23.3 Risk-Based Device Inspections and Statistical Sampling

Different approaches have been used to reduce the amount of resources needed to inspect measuring instruments without performing 100 % device inspection on an annual basis. A state may perform a test at each service station on an annual basis, but the inspectors do not necessarily test every dispenser at the station. The state would provide a list of all of the dispensers at the service stations that an official is to inspect.

For each service station, specific dispensers (meters) would be identified for testing in a three-step process. First, a certain percentage of meters would be tested. If any of the meters fail the inspection, then the inspector would test a larger number of meters that are also identified on the inspection sheet that was sent to the inspector by the state office. If a certain number of dispensers in the larger sample fail the inspection, then the third step is for the inspector to test all of the dispensers (meters) in the station. In this case, all businesses would have some measuring instruments inspected, but not necessarily all of the instruments. The effect of this approach is that all businesses are aware of weights and measures enforcement because they receive an annual inspection.

Another approach to statistical sampling is to apply the concept of risk-based sampling. In this approach, several factors may influence which businesses or measuring instruments are inspected. For example, if a particular business has a history of high rejection rates, the business may be scheduled for more frequent inspections than the stores (or a chain of stores) with low rejection rates. Not all businesses may be inspected annually. Another factor may be to focus most of the inspection on the measuring instruments that have the highest volume of transactions versus the instruments with a lower volume of transactions. Yet another factor may be the number and frequency of consumer complaints in a region or within a specific category of device or business. A certain number of measuring instruments should be tested on a random basis to ensure that all measuring instruments in the store are being maintained in an accurate condition.

Many variations of risk-based statistical sampling may be used. The important characteristic is that the database, the compliance history of the business, and human judgment on the economic importance of certain measuring instruments over others are used to focus inspections on where problems exist, on the highest volume devices, and those of greatest economic impact. One can expect that the risk-based inspections will probably result in a higher rate of noncompliance, because problem areas are targeted for inspection. This result should be considered a success, because resources are being used more effectively to correct problem areas. A random market survey should be conducted periodically to determine the overall compliance rate for a particular type of measuring instrument or a particular segment of the marketplace.

This approach works best under the following conditions:

1. The weights and measures agency has a database of all the businesses and measuring instruments in use within the jurisdiction.

2. An inspection itinerary is developed in advance for the weights and measures inspector.

3. The compliance level for the type of measuring instrument targeted for statistical sampling is already at a high level as determined by a more extensive inspection program.

4. The sampling program is statistically sound and incorporates risk-based inspection criteria.

5. Businesses with measuring instruments routinely obtain service from private service companies to test and maintain their instruments in compliance with weights and measures requirements.

6. The state uses periodic marketplace surveys to independently sample all measuring instruments of a given type that are in use within the jurisdiction and the results of the marketplace surveys are consistent with the results of the risk-based sampling program. The level of compliance must continue to be high over time.

23.4 Integrating Government and Private Sector Inspections

The pressure to reduce spending at all levels of government is ongoing, and competition for general revenue funds is intense. Unfortunately, in many instances, weights and measures programs are unable to compete effectively with the many other demands for these funds, including programs in health, public safety, education, and environment.

As a result, many weights and measures programs have experienced decreases in their budgets, which has subsequently weakened the infrastructure of the commercial measurement system. Weights and measures programs must continually seek better and more efficient and effective ways to monitor the commercial measurement system and ensure fair competition and consumer protection. Some states legislatures have moved to "privatize" weights and measures inspection and enforcement responsibilities, that is, to turn over government inspection and enforcement to the private sector.

In reality, weights and measures programs must explore alternatives on how to effectively monitor the commercial measurement system with fewer resources. A state may have a program that incorporates the work done by service companies with government follow-up inspections and extensive testing in the marketplace. This program may be applied to one segment of a service industry, but the concepts apply to any companies that provide service and repair for measuring instruments. Elected officials must understand that an extensive management and auditing activity is needed if weights and measures is to incorporate private service companies into the regulatory process of overseeing the marketplace.

When government and private sector inspections are integrated, states use the information from service companies to supplement their own inspection records. The oversight program operated by the state is essential to ensure that the information provided by the service companies is valid and that there is fair competition among the service companies. It is critical that the devices tested by the service companies are accurate and comply with weights and measures laws and technical regulations. This determination is made by reviewing the paperwork submitted by the service companies, requiring annual training in the state weights and measures requirements, observing the test procedures used by the service technicians, and conducting follow-up testing of devices that were checked by the service companies to verify that the devices are performing consistent with the information reported by the service company.

Since service companies are profit driven, some companies and service technicians may try to abbreviate the test procedure to shorten the time needed to provide the service. Remember too, that service companies are often pressured by their customers to reduce costs or face losing business, so it's important to identify the driving force behind the failure of a service agency to conduct thorough examinations. Adequate requirements must be in place to ensure that service companies and their technicians are knowledgeable in the state weights and measures requirements, and that the final results of the service work leave the device in compliance with those requirements.

The integration of government and private sector inspections works best under the following conditions.

1. The state has the auditing and investigation resources to operate an effective oversight program to monitor the performance of service companies, including a mechanism for device owners to report problems.

2. Effective penalties are available and are used when private service companies are not meeting acceptable standards.

3. The service industry is sufficiently developed so there is rigorous competition between service companies, and the companies have knowledgeable, competent, and reliable technicians.

4. A training and testing program is in place for service agents.

5. The state weights and measures officials perform follow-up inspections that include a review of service reports to determine if service company performance meets acceptable standards.

6. The cost of managing the oversight program is less than the government performing the inspections with government employees.

7. The weights and measures agency has a database to efficiently input the inspection results with their own inspection results and uses the information from both sources to maintain the infrastructure of the commercial measurement system.

23.5   Delegating Inspection Responsibilities to Private Companies

A few states have delegated the inspection and test responsibilities for some types of measuring instruments to private service companies. The difference between delegating the inspection responsibilities to the private sector and integrating the inspection results from private service companies into the state weights and measures inspection records is the extent to which field inspections continue to be conducted by the government weights and measures inspectors. The expectation when integrating inspections is that the state is still conducting annual tests on at least 50 % or more of the measuring instruments in the field. This implies that the state has adequate oversight of the commercial measurement system. When inspections are delegated to the private sector, the state may only conduct follow-up inspections on a small percentage of devices in the field.

Whether delegation or integration is used, , the state weights and measures program should conduct periodic random market surveys of measuring instruments in the field to assess the accuracy of measuring instruments to see if the random sample results agree with what is reported through the routine inspection program. Any discrepancies in test results should be investigated and corrective action taken to ensure uniformity and consistency in test results.

The delegation of field testing and inspections from a government agency to non-government bodies, such as private service companies, can take many forms. Two forms of delegation are the assignment of inspection/regulatory responsibility to service companies and witnessed testing. Assignment of inspection/regulatory responsibility to service companies requires that the state weights and measures regulatory authority exercise considerable oversight and periodic field inspections of its own to ensure that service companies are fulfilling their responsibility, following the correct test procedures, making the appropriate repairs, properly sealing devices, and properly reporting test results.

However, experience has shown that service companies often have a conflict of interest of their "for profit" service and repair business and their assigned inspection/regulatory responsibilities. Consequently, the assignment of regulatory authority and inspections to private companies must be carefully considered, carefully implemented, and rigorously overseen by the regulatory agency, because many problems have been reported over the years by jurisdictions that have implemented these types of programs. In general, the assignment of inspection/enforcement to private companies should be avoided.

One major criticism of this approach is that many view the service companies as having a conflict of interest when given a "regulatory" responsibility to inspect and test measuring instruments and then they have the private sector responsibility to service the instruments and put them back into commercial service. Another criticism is that the government is abdicating its responsibility to oversee and regulate the commercial measurement system. To make this approach work, some states have required the owners of measuring instruments to have them tested annually.

Again, it is critical that an adequate government oversight program be in place to make this approach work effectively. The service technicians essentially become unsupervised representatives of the state. The oversight program is needed to ensure that service companies are using the correct test procedures, doing the work and achieving the quality of work that they report. It is essential for fair competition among service companies that all service technicians are performing work that meets the minimum acceptable standards for the state weights and measures program. Adequate requirements must be in place to ensure that service companies and their technicians are knowledgeable in the state weights and measures requirements, the device is in compliance with those requirements after repair and that the performance of the device is consistent with the performance reported by the service company.

Some of the oversight activities that should be carried out are listed below.

- Ensure that the representatives:

  o have a thorough understanding of the weights and measures regulatory requirements;

  o have adequate and certified field standards;

  o use the proper test procedures; and

  o Ensure that device specifications and performance requirements are properly applied;

- Conduct periodic direct observation of the conduct of the work performed by service technicians to ensure that proper inspections are being conducted;

- Have all inspection reports submitted to the weights and measures director for inclusion in a database of inspection results;

- Analyze and compare the results of different private companies performing regulatory inspections to ensure consistency and to look for any aberrations of results; and

- Schedule periodic follow-up inspections by government weights and measures inspectors to verify the test results on test reports submitted by the private companies.

Some jurisdictions do not have funds for an adequate oversight program when inspection responsibilities are turned over to the private sector. A few policymakers may mistakenly support "privatization" of weights and measures inspections as a way to cut government expenditures without fully understanding the economic consequences this may have on consumers and businesses. Experience with failed efforts in the past and the experiences with some current efforts as reported by weights and measures directors at NIST workshops point out the many problems associated with "privatization" efforts. The types of problems that have been reported are:

- Service technicians are not adequately trained in weights and measures requirements;

- Service companies have falsified reports;

- Companies do not follow the specified test procedure and run abbreviated or incomplete tests;

- State inspectors find missing or broken seals on adjustment components during follow-up inspections;

- State inspectors find sealed devices that are outside of tolerance limits; and

- Service companies report that calibrations have been performed, but no adjustments have been made to the measuring instruments.

A considerable level of oversight is needed for this approach, because the state inspectors are not conducting nearly the number of tests as they would normally perform. Some states believe that the oversight program requires more resources than would be required if the state inspectors performed all of the tests, because of the extensive oversight needed to monitor the large number of nongovernment people performing the same number of tests.

The delegation of inspection responsibilities to the private sector works best under the following conditions:

1. The state has an effective oversight program to monitor the performance of service companies.

2. Effective penalties are available and used when private service companies are not meeting acceptable standards.

3. The service industry is sufficiently developed so companies have knowledgeable, competent, and reliable technicians.

4. Training and testing program is in place to assure competency.

5. The state weights and measures officials perform follow-up inspections and find that service company performance meets acceptable standards.

6. The cost of managing the oversight program is less than the government performing the inspections with government employees.

7. The weights and measures agency has a database to efficiently input the inspection results with their own inspection results and uses the information from both sources to maintain the infrastructure of the commercial measurement system.

23.6   Witnessed Testing

In witnessed testing, a weights and measures regulatory official goes to a business location and observes the testing done by a private service company or by a representative of the business itself. Witnessed testing is frequently done for the test of vehicle scales and static and in-motion railway track scales and belt-conveyor scales, because the preparation time and arrangements for the tests are time consuming and expensive. For example, a state program may not have a test car to test static-weighing railway track scales The most appropriate test car is one carrying at least 100 000 lb of test weights, including an electric-powered weight cart that moves the weights across the scale and also serves as a field standard test weight. The weights and cart are transported in a specially designed railroad car and moved by the railroads as their schedules permit.

Moving the test car from one railway scale to another may take days or weeks, and the arrival dates and times are frequently uncertain until just before the test car arrives. When testing coupled- or uncoupled-in-motion railway track scales, empty and loaded railroad cars must be weighed on a reference scale, which may take a great deal of time and must be coordinated with the railroads, which provide the labor, the locomotive, and the railcars to be used in the tests. Similarly, when large belt-conveyor scales are tested, frequently empty and loaded railroad cars must be weighed on a reference scale so that known amounts of material can be passed over the belt-conveyor scale. Several tests are usually required on belt-conveyor scales, especially when adjustments must be made. Because of the cost and time required to arrange and conduct these tests, only a few weights and measures programs witness the tests.

Witnessed testing may be conducted on other types of measuring instruments; however, for smaller capacity scales and meters, the weights and measures official often has adequate field standards with valid certifications to test the smaller instruments. Generally, fewer questions exist regarding the validity of the standards used in tests when the standards are under the exclusive control of the weights and measure regulatory official at all times.

Witnessed testing on the part of a government inspector has advantages and disadvantages. The advantages are that the inspector does not have to own or bring the test equipment to the test site and the inspector can observe the test procedure used by the private service company. The questions that must be addressed before witnessed testing is used as part of a regulatory program include the following:

- When the service company is not using the accepted test procedure, is it appropriate for the weights and measures inspector to require that the proper test procedure be used?

- Is it appropriate that the business that owns the scale or metering device be charged for the additional time required to conduct a regulatory inspection and test when the service company normally would not perform all these tests to service and adjust the device?

- Who decides if additional testing is necessary, for example, to verify device performance as being in or out of tolerance or after an adjustment?

24.0   Record System

An efficient record system of inspections and enforcement actions is essential. The information in the records must be analyzed and used as a tool to justify the program, to guide the allocation of resources, to monitor compliance levels in different segments of the commercial marketplace, to monitor compliance levels for retail chains, manufacturers and models of measuring instruments and consumer product packagers, and to monitor the performance of service companies of measuring instruments. Recordkeeping systems points to consider:

- The law usually specifies that records shall be kept of all standards and equipment and of all official acts, although the details of the record system are not generally prescribed.

- The recorded data should be sufficiently detailed in character so that it can fully answer all relevant questions

- The system of indexing and filing the records should be simple and effective, so that desired information may be located quickly when it is needed.

- Because conditions differ among the hundreds of weights and measures jurisdictions throughout the country, no single system can be recommended for all cases.

Today, most recordkeeping systems for weights and measures are computerized. The challenge now is how best to get inspection results into the database and then how to extract meaningful information from the database. Collecting data without analyzing the data is a waste of valuable effort and information. It is important to use the data to provide insights into the activities and outcomes of weights and measures programs.

The information to be collected and stored in a database should be determined by the type of information and reports that are needed and desired for the analyzed data. The following information from the inspections of measuring instruments is routinely collected by weights and measures programs:

- Business identification number if assigned by the weights and measures program;

- Business name, address and telephone number;

- Owner name, address and telephone number;

- License or business registration number (if applicable);

- Name of inspector who conducted the inspection;

- Device type according to categories;

- Device manufacturer;

- Model number;

- Serial number;

- Inspection date;

- Action code for pass or reason for rejection based on noncompliance with Handbook 44; and

- Device error magnitude and direction (for example, for meters, fast flow and slow flow errors; for scales, test that fails, e.g., increasing load, decreasing load, shift test, return to zero, and the maximum error as a function of the tolerance or test load, etc.).

The following information is routinely collected for net content package inspections:

- Business identification number if assigned by the weights and measures program for the location where the packages were tested;

- Business name, address and telephone number;

- Owner name, address and telephone number;

- License or business registration number (if applicable);

- Name of inspector who conducted the inspection;

- Product identity;

- Manufacturer of the packaged product;

- Package size (net content);

- Lot size;

- Lot code(s) and, if applicable, number for the U.S. Department of Agriculture seal;

- Inspection date;

- Audit test or compliance test;

- UPC code;

- Average error; and

- Action code for pass or reason for rejection based on noncompliance with Handbook 133, i.e., failed average net weight or failed maximum allowable variations for individual packages or both.

25.0 Analysis of Data

The following actions are recommended for net content inspections:

- Package inspection data are analyzed by manufacturer and by types of product to establish an estimate of the manufacturers packaging practices.

- Follow-up net content inspections are conducted at distribution warehouses or at the manufacturer's plant to obtain packaging characteristics on larger lot sizes.

- Short-weight packaging information is shared with states in which the manufacturer of a product has its manufacturing plant so inspections may be performed at the plant to obtain packaging characteristics on larger lot sizes.

- Moisture loss is considered for all short-weight packages where appropriate.

- Instances of observed short weight or measure of products falling under USDA and FDA are communicated to the appropriate agency.

25.1   Examples of Analysis for Retail Motor-Fuel Dispensers[17]

Once inspection data is collected, it must be analyzed to gain the greatest benefit of the data. The data is helpful to assess and demonstrate the effectiveness of the weights and measures program, to identify trends in inspection results, to assess the compliance performance of individual businesses, to assess the performance of measuring instruments provided by different manufacturers, and to assess inspector performance.

Figure 5 shows retail motor fuel dispenser flow delivery error rates for all meters in 2006 (a total of 20,036 meters). Approximately the same number of meters were out of compliance for over delivery as for under delivery. The largest single group showed a zero error rate. The data demonstrates a normal distribution around the zero error rate, which is expected in a well functioning system.

---

[17] Appreciation is expressed to the Nebraska Bureau of Weights and Measures for the use of the graphs below as examples of their analysis of inspection results for retail motor-fuel dispensers.

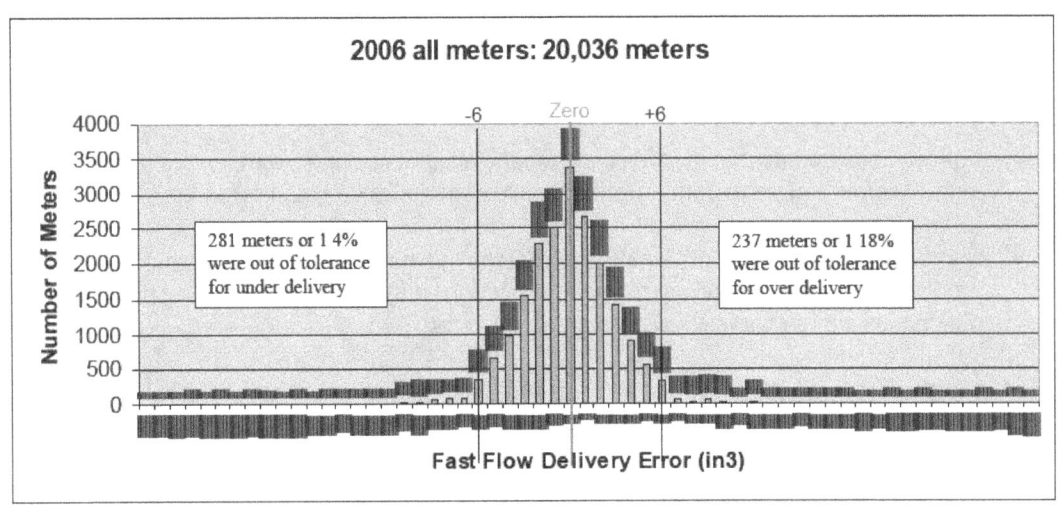

Figure 5. Retail Motor Fuel Dispenser Flow Deliver Error Rate

Figure 6 shows compliance rates by manufacturer. Only one manufacturer had a compliance rate under 80 %, and four were at the 100 % level. Based on this data, a program conducting less that 100 % inspections may focus future inspections on devices from specific manufacturers.

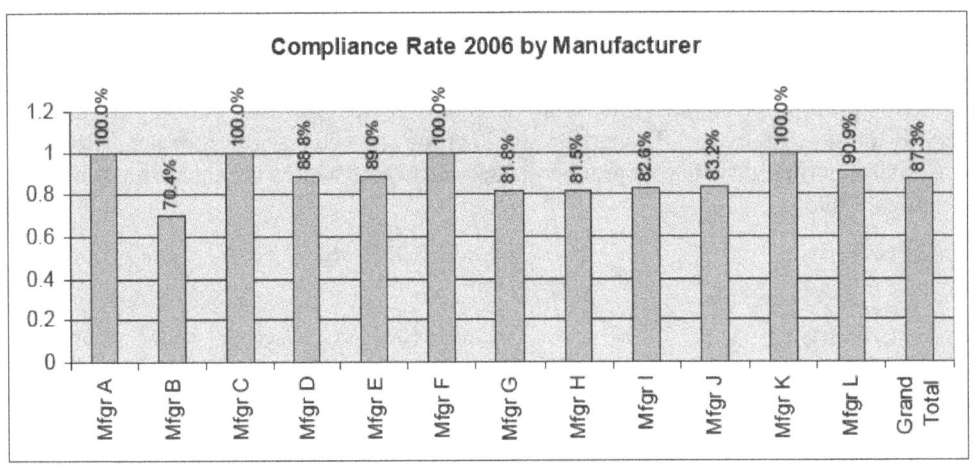

Figure 6. Retail Motor Fuel Dispenser Compliance Rate by Manufacturer, 2006

Figure 7 shows rejection codes for seven different manufacturers. The graph indicates that code 8 (indication and recording), code 65 (pump plus error), and code 80 (inoperable) were the three most frequent reasons for rejection. One manufacturer also had a high rejection rate for code 90 (predominance error).

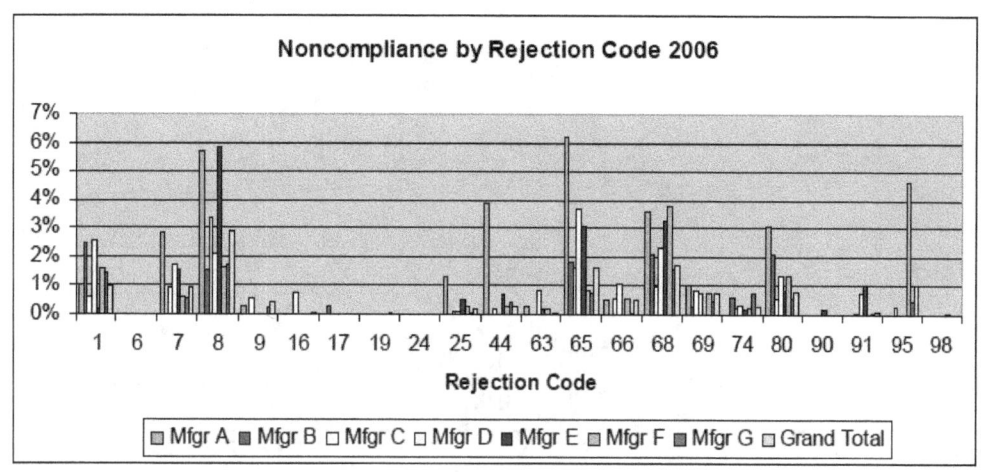

Figure 7. Noncompliance by Rejection Code, 2006

Table 3 below lists all the rejection codes that were used in this analysis.

Table 3. Retail Motor Fuel Dispenser Rejection Codes

Rejection Codes	
01 Unable to Test	63 Measuring Elements
06 Installation	65 Pump Plus Error
07 Customer Readability	66 Discharge Hose
08 Indication and Recording	68 Pump Minus Error
09 Sealing	69 Marking Requirements
16 Environmental Factors	74 Nozzle
17 Repeatability	80 Inoperable
19 Air Eliminator	90 Stop Use Tag
25 Computer Jump	91 Interlock
44 Price Computation	95 Predominance Error
	98 Temporary Approval

The two graphs in Figure 8 show fast flow errors by manufacturer and by owner. In each case the bell curve is skewed with a disproportionate number of negative error readings. In the case of owner ABC, he could be cited for using tolerances to his advantage. Since the use of tolerances is subjective, capturing and maintaining this type of data can be very helpful to the regulatory official, giving solid evidence for any enforcement action.

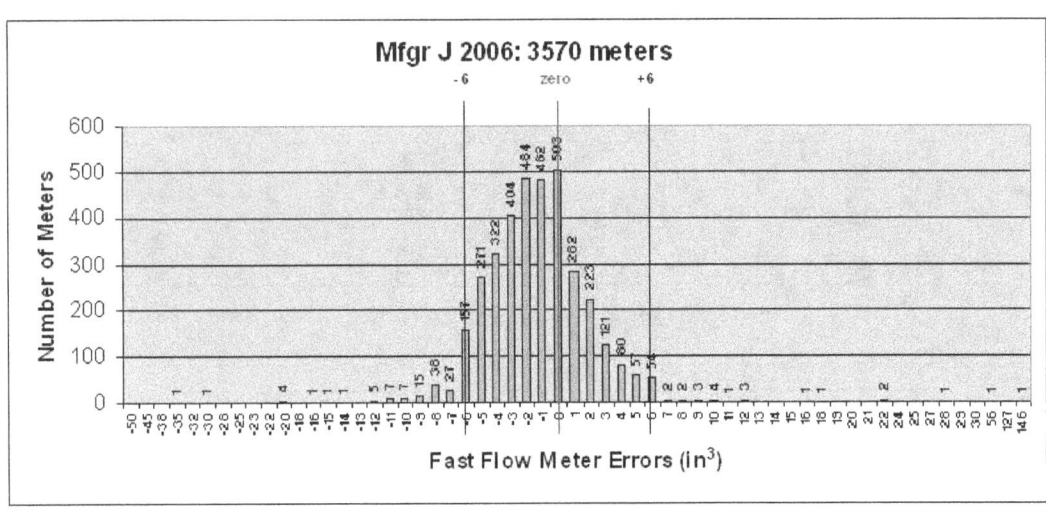

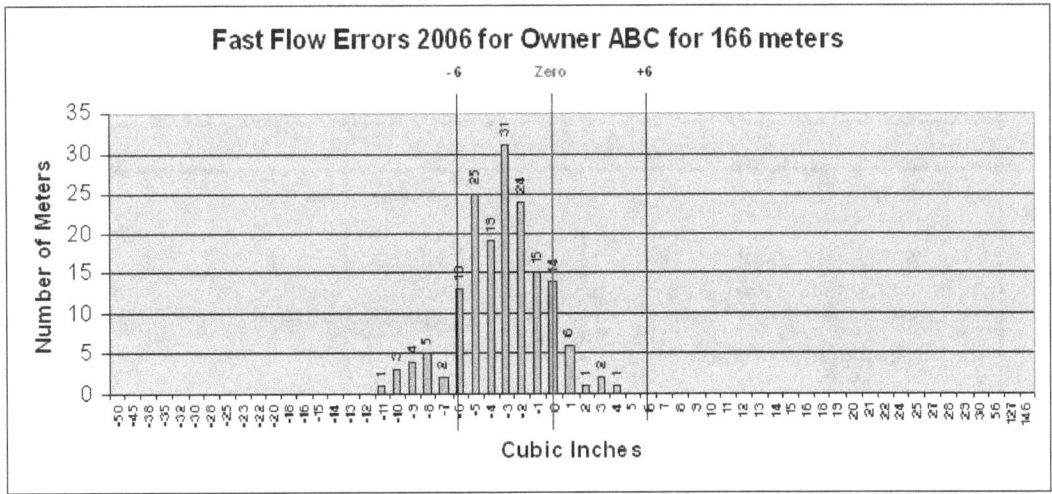

Figure 8. Retail Motor Fuel Fast Flow Error Rates, 2006

26.0 Industry Relations

The retailers, the manufacturers of measuring instruments, and the packagers of consumer products are the major sources of information on the economic and marketing impact of weights and measures regulations. Communicating with and through industry associations is an efficient means to get information to stakeholders. It is recommended that:

- Weights and measures officials maintain cordial but businesslike relationships with representatives of companies engaged in manufacturing, selling, and servicing commercial devices.

- The experience and ideas of officials are valuable to manufacturers.

- Similarly, officials may learn about details relative to the construction, adjustment and operation of the devices from the manufacturer.

- Cooperation should be encouraged in order to promote continuous improvement throughout the weights and measures community.

Weights and measures programs should have activities that reach out to the various stakeholders whenever changes are being made to the regulatory program. Appropriate outreach can accomplish the following:

- Inform and educate stakeholders regarding changes to regulations;

- Provide answers to questions regarding enforcement; and

- Communicate expectations for compliance with regulations.

The goal is to help industry to develop the capability to comply with weights and measures requirements rather than to rely purely on enforcement inspections to achieve compliance.

Because many stores are part of regional or national chains that may operate many stores in a state, some weights and measures programs operate corporate education programs. Through corporate education programs, weights and measures officials can inform the managers of several stores simultaneously about a program and seek the cooperation and assistance of these managers to achieve compliance. Corporate education programs are win-win situations that facilitate communication and compliance.

Many weights and measures programs report that their strongest support comes from the industries with which they have frequent communication, because of the cooperative nature of these interactions. When there is mutual respect between regulated industry and the regulator— that is, when the regulated industry believes he is treated fairly and that the weights and measures program is looking out for his best interests—he is motivated to defend the weights and measures program at the legislature or in technical committees. This is because he sees the value in a continued robust program. Thus, industry support and effective annual reporting go hand in hand in demonstrating the worth of weights and measures enforcement.

27.0   Benchmarking

Weights and measures directors frequently seek a basis for comparing the effectiveness of their program with those of other jurisdictions. This practice is very desirable, because the director will have an outside basis for comparison of compliance rates in the many different disciplines and the productivity of inspectors. However, in some cases comparison is difficult because inspection activities vary from state to state. Possible variations include the following:

- Some states may only test retail fuel dispensers for accuracy, but do very little inspection regarding the specifications of the installed dispensers.

- Some states may test measuring instruments annually and other states may test only every two or three years.

- Some programs may only conduct normal tests on liquid meters, while others will conduct both the normal test and special tests.

- Some states may test vehicle scales by placing test weights on only one end of the scale (or worse yet, just driving the test unit across the scale and using it as a "rolling standard"), whereas another state may have weight carts and they are able to test every section of a vehicle scale.

- States may use different amounts of test weights, so the rejection rates may not be comparable.

- States may perform strain load tests in addition to the increasing and decreasing load tests on large capacity scales, which could result in different rejection rates.

Each state may have different capacity categories for small, medium and large capacity scales or meters, which means that the rejection rates for these instrument categories may not be directly comparable.

Despite these difficulties, weights and measures programs should benchmark their programs and inspection results with those of similar programs. Furthermore, all weights and measures programs should work to find a way to extract inspection results from their databases so they can compare inspection results for common types of measuring or inspection activities.

The NCWM has recommended a categorization scheme for measuring instruments (see Table 4) that jurisdictions are encouraged to use, either by revising their current device categories or supplementing their current categories with an additional category structure in their database that will allow comparison on a national basis. The device registration programs of many states are based on different categories of devices, so it is difficult to change the device registration categories. However, by adding another category structure to their database, states will ultimately be able to compare inspection results across the country.

Table 4. NCWM Device Category Codes

DEVICE CODE	CATEGORY	CAPACITY	EXAMPLES
	**DEVICE CATEGORY CODES**		

DEVICE CODE	CATEGORY	CAPACITY	EXAMPLES
SP	Scale, Precision	< 5 g scale division	jewelry, prescription scales
SS	Scale, Small	< 300 lb	retail computing scales
SM	Scale, Medium	301 to 5 000 lb	dormant, platform scales
SL	Scale, Large	> 5 001 lb	livestock, recycler scales, hopper scales, belt conveyor
SV	Scale, Vehicle	> 40 000 lb	vehicle, railway track scales
MS	Meter, Small	< 30 gpm[1]	retail motor-fuel dispensers
MM	Meter, Medium	30 to 200 gpm	vehicle-tank meters
ML	Meter, Large	> 200 gpm	agri-chemical meters, bulk oil meters, loading rack meters
MF	Meter, Mass Flow	All	heated tanks of corn syrup (soft drinks)
MW	Meter, Water	All	water sub-meters for mobile homes & apartments
MG	Meter, LPG	All	propane sales
MT	Meter, Taxi	All	taximeters
DT	Device, Timing	All	clocks in parking garages
DL	Device, Length Measuring	All	cordage meters
GM	Grain Moisture Meter	All	
GA	Grain Analyzer	All	
MD	Multiple Dimension Measuring Device	All	
MC	Meter, Cryogenic	All	

[1] Retail motor-fuel dispenser counts are based on meters.

An important method that can be used for benchmarking, despite differences in categorization in databases, is to conduct national marketplace surveys in different weights and measures disciplines.

Market surveys should be conducted regularly and in different inspection areas. The surveys should target both common inspection areas and areas where weights and measures inspections are limited. If compliance rates are low in areas of significant economic importance and weights and measures oversight has been limited, resources should be shifted to improve the accuracy of transactions in the segments of commerce with low compliance rates.

When a state participates in a market survey, the weights and measures program should use the survey results to compare with the compliance rates obtained in the routine inspection program. If there are differences in the compliance rates from the market survey and the routine inspection program, then the reasons for the differences should be identified and, if necessary, actions taken to resolve the discrepancies.

Market surveys must be based upon commonly used and accepted procedures so that the results from each state can be compared to others. Inspectors should use common report forms, traceable standards and certified test equipment in all market surveys, as well in their routine inspection activities. The analysis of market surveys should be unbiased and based on statistically sound methods.

Before a new inspection program is initiated in an area where there may have been little or no weights and measures inspections before, a random sample of measuring instruments or businesses should be inspected to establish a baseline compliance rate to see if and how quickly accuracy in transactions improve as a result of the new inspection program. Improvements in compliance rates are valuable data to demonstrate the value and effectiveness of weights and measures programs.

28.0   Conclusions

Transactions involving weights and measures touch upon virtually every aspect of economic life in the United States, affecting about half of the gross domestic product. Ensuring fair and accurate trade is critical. Weights and measures officials serve a vital role in this process, as an impartial third party between buyers and sellers.

The weights and measures community includes a complex infrastructure of suppliers, manufacturers, government officials, legislators, and consumers. When all the participants cooperate, the commercial measurement system works efficiently and effectively, which results in a high level of confidence on the part of both buyers and sellers.

A key component of this collaboration is the open exchange of information. Weights and measures officials should educate businesses about requirements for inspection and compliance. Consumers can inform officials when irregularities occur in the sale of products so that fraudulent or erroneous transactions can be corrected, and if necessary, penalties imposed. Uniformity in laws and procedures helps ensure consistent outcomes.

Funding a weights and measures program poses a continual challenge. Because the activities of weights and measures are not highly visible and their value is not always recognized, legislators may be reluctant to fund the programs. A variety of options are available for generating revenue to support weights and measures programs, including funding from fuel taxes, device registration, licensing, and inspection fees.

The technical requirements placed on weights and measures officials are continually increasing. When possible, inspectors should be able to devote full time attention to their job rather than having to take on other duties. Ongoing personnel development is vital to maintaining a skilled workforce. Weights and measures employees should have opportunities for training, attendance at conferences, and interaction with experts in the field in order to continually increase their knowledge.